Early ac

"The less dust,

"An impressive view into the vast landscape of the Shamatha Project, this book is a rich account of the practices and outcomes from this pioneering endeavor of mapping meditative experience."
– Joan Halifax Roshi, Founding Abbot, Upaya Zen Center, author of *Being with Dying*.

"What I personally find so compelling about this book is its accessibility. By the warmth and honesty of her writing, Adeline van Waning gives one the assurance of a friend who walks beside you, telling you how it was for her and her colleagues as they progressed through their 'expedition,' their three month Shamatha Project. She presents the practice guidance that she received in a way that it may offer a valuable path for all readers."
– Sherry Ruth Anderson PhD, Ridhwan teacher, co-author of *The Feminine Face of God*, author of *Ripening Time*.

"In this volume Adeline van Waning admirably brings to bear her professional training and experience as a psychiatrist together with her knowledge and experience as a meditator to explain the nature and significance of these practices from both Buddhist and scientific perspectives ... With her exceptional background as a scientist and as a meditator, Dr. van Waning bridges the gap between third person and first person methodologies, showing how each one can complement the other. This, clearly, is the way forward if we are to seek the most complete understanding of the mind and consciousness."
– B. Alan Wallace PhD, Buddhist meditation teacher, scholar, Director of the Santa Barbara Institute for Consciousness Studies, author of *The Attention Revolution*, and *Mind in the Balance*. From: Prologue.

"The many examples of shamatha meditation guidance as presented in this book, including attentional practices, the Four Qualities of the Heart and *Tonglen*, can be very useful for students on various paths in navigating their own journey with meditation."
– Lama Palden Drolma, Founder and Resident Teacher of Sukhasiddhi Foundation.

"This is an informative and engaging work of a very high standard. It will appeal both to Western Buddhists interested in meditation and scientists interested in the measurable effects of meditation and the implications of this for understanding the brain and consciousness. This very systematic, well structured and thoughtful study is a valuable description, contextualization and analysis of a three month meditation 'expedition' led by B. Alan Wallace. It focuses on *shamatha* meditation as practiced in the Tibetan tradition, accompanied by scientific assessment of effects on participants."
– Peter Harvey PhD, Professor Emeritus of Buddhist Studies, University of Sunderland, UK, Editor, *Buddhist Studies Review.*

Endorsement by a co-research subject in the Shamatha Project:
"I am both gratified and relieved that Adeline has written this ambitious book: gratified because our extraordinary opportunity and experiences in this project simply needed to be documented, and relieved because she is exactly the right person for the job. The Shamatha Project set a new standard of rigor in design and methodology for studies of meditation, and this book reflects that standard in its thoroughness and loving rendering. It provides an insider's view of the gratitude and life-changing shifts we research subjects enjoyed as we daily received impeccable, authentic teachings and then meditated for long hours in an idyllic setting high in the Rocky Mountains, all the while certain that we were simultaneously contributing to science, to Dharma, and to the cultivation of our own hearts and minds. Adeline's intelligent and thoughtful psychological and philosophical contextualization of her personal experiences

makes this book appealing to those interested in meditation, Dharma, contemplative neuroscience, and the many hybrid and integrative disciplines arising from them. May this virtuous effort by my favorite 'Shamatha Buddy' enrich your understanding and commitment to your own path of virtue, joy, and liberation."
– Jim Cahill, BCB, Developer, Mindfulness-based Biofeedback Therapy™

"The Less Dust, the More Trust"

Participating in
The Shamatha Project
Meditation and Science

"The Less Dust, the More Trust"

Participating in
The Shamatha Project
Meditation and Science

Adeline van Waning

MD PhD Psychiatry

MA Buddhist Studies

Winchester, UK
Washington, USA

First published by Mantra Books, 2014
Mantra Books is an imprint of John Hunt Publishing Ltd., Laurel House, Station Approach, Alresford, Hants, SO24 9JH, UK
office1@jhpbooks.net
www.johnhuntpublishing.com
www.mantra-books.net

For distributor details and how to order please visit the 'Ordering' section on our website.

ISBN: 978 1 78099 948 7

A CIP catalogue record for this book is available from the British Library.

Design: Stuart Davies
www.stuartdaviesart.com

Printed and bound by CPI Group (UK) Ltd, Croydon, CR0 4YY

CONTENTS

List of guided meditations

List of illustrations

(All illustrations by Adeline van Waning)

Prologue

by B. Alan Wallace PhD

The meditative practices that were central to the scientific study known as the Shamatha Project consist of two kinds: (1) the shamatha practices of mindfulness of breathing, settling the mind in its natural state, and shamatha without a sign; and (2) the cultivation of four sublime qualities of the heart known in Buddhism as the four immeasurables, namely loving-kindness, compassion, empathetic joy, and equanimity. The first set of practices is explicitly designed to develop attentional balance through the cultivation of relaxation, stability, and vividness, qualities that are then used to introspectively explore the nature of the mind and its potentials. The second set is for the sake of developing a greater sense of kindness and acceptance toward oneself and others, which has a deep impact on one's values, ideals, and emotional balance.

In this volume Adeline van Waning admirably brings to bear her professional training and experience as a psychiatrist together with her knowledge and experience as a meditator to explain the nature and significance of these practices from both Buddhist and scientific perspectives. Since the turn of the century, a rapidly growing number of scientific studies have revealed the health benefits of various kinds of mindfulness-based meditation. Brain scans, EEG measurements, behavioral studies, and questionnaires have shown the influence of meditation on the brain and behavior, which in the minds of many people lends some degree of credibility to the practice of meditation. In the overwhelming majority of such studies, those who conduct and report on the research are professionally trained scientists, intent on applying objective measures to understanding the nature and effects of meditation. In contrast, the meditators are treated as subjects in these studies, similar to human and non-human subjects in other psychological and neuroscientific kinds of research. So their identities are almost invariably ignored in scientific reports on meditation, and all discoveries pertaining to meditation are claimed by the scientists, who in many cases have little or no meditative experience. Consequently, whatever discoveries about the nature of the mind may have been made by the meditators themselves are generally overlooked in scientific papers, presumably because they are not deemed "objective" and are therefore not "scientific."

This bias for objective, third person evidence over subjective, first person experience is problematic when it comes to understanding the nature of mental processes and states of consciousness, all of which are undetectable by all objective systems of measurement. In contrast, on the basis of the development of refined attention skills using the above methods of shamatha meditation, one gains an increasing ability to observe a widening array of mental processes and states of consciousness. By so doing, one may make discoveries about the mind that are inaccessible to third person methods of observation. With her exceptional background as a scientist and as a meditator, Dr. van Waning bridges this gap between third person and first person methodologies, showing how each one can complement the other. This, clearly, is the way forward if we are to seek the most complete understanding of the mind and consciousness.

Among the three methods of shamatha explained in this book, "settling the mind in its natural state" is especially oriented toward making first person discoveries about the unique qualities of one's own mind and about the nature and potentials of the mind in general. Bearing many qualities in common with "insight meditation" practices commonly taught in the Theravada and Zen traditions of Buddhism, it enables one to make internally "objective" observations of the origination, nature, and dissolution of discursive thoughts, desires, emotions, and other mental processes. In this way, such experiential inquiry has great epistemic value for understanding the mind firsthand. In addition, however, the practice of maintaining clear, nonreactive awareness of such mental events also has great therapeutic value, which is clearly explained in this volume by a mental health professional. So the epistemic and pragmatic significance of this and other shamatha methods are deeply integrated: knowing thyself is integral to healing thyself. There is a similar synergy between the practices of shamatha and the four immeasurables, each one deepening and enriching the other.

While modern popularizers of yoga and meditation often teach various methods as stand-alone techniques, independent of any theory, values, or lifestyle, this reductionist approach is alien to the Buddhist tradition and all other great contemplative traditions of the world. If one adheres to a materialistic worldview, believing that everything in the universe, including all living organisms and states of consciousness, can be thoroughly understood solely as emergent properties of matter, this must have a direct impact on one's values and priorities. If one believes that only matter and its emergent properties are real, those are the only things one will value, and the only kinds of happiness one will seek are stimulus-driven, hedonic pleasures, arising from interactions of matter and energy. Moreover, if one's values are thoroughly materialistic and hedonic, this will inevitably result in a consumer-driven way of life bent on material acquisition and consumption and the pursuit of hedonic pleasures.

Buddhist meditation, in contrast, is embedded in a worldview that embraces both physical and nonphysical elements of the natural world. Within the Buddhist Eightfold Noble Path, authentic mindfulness and concentration arise only in conjunction with an authentic worldview—not subject to the limitations of materialism—and with an authentic aspiration oriented toward genuine happiness, which arises from ethics, mental balance, and wisdom.

The origins, nature, and potentials of consciousness, together with the nature and means of realizing genuine happiness, are of the utmost importance, especially in today's world, in which the devastating effects of unbridled materialism are wreaking havoc on modern society and the natural environment. Materialistic assumptions about human nature continue to hamper open-minded inquiry into the relation between the body and mind, including how consciousness first emerges in a human fetus and what happens to it at death. Materialists assume it first emerges from complex interactions of neurons and it simply disappears at death, but they have never scientifically demonstrated the truth of their beliefs. Contemplatives from multiple traditions East and West reject that assumption, but the first person discoveries on which they base their conclusions have yet to be taken seriously by the scientific community.

As an analogy, even after Copernicus presented his brilliant heliocentric theory of the movements of the planets around the sun, medieval scholastics continued to cling to their belief that the sun and planets orbited around the earth. Both the heliocentric and geocentric views accounted for the appearances to the naked eye of the relative movements of these celestial bodies. It was only when Galileo refined the telescope as an instrument for making precise observations of the sun, moon, and planets, that he was able to discover the phases of Venus, which provided irrefutable evidence that the medieval geocentric view was invalid.

Nowadays, learned scientists and theologians continue to debate about the fate of human consciousness after death, with each group adhering to their own assumptions, without being able to point to evidence that settles the issue for all intelligent, open-minded seekers of truth. According to Buddhism, the achievement of highly refined states of focused attention, trained inwardly, transcending the limitations of the normal human psyche, sheds light on dimensions of consciousness that are not contingent upon the brain. If this discovery is valid and can be replicated by anyone with sufficient contemplative training – regardless of their metaphysical beliefs – this will shift the modern understanding of the mind from a "materiocentric" to an "empiricocentric" view. This will herald the first scientific revolution in the mind sciences, in which experience will once again triumph over dogma, and antiquated metaphysical beliefs about the nature and potentials of consciousness will be defeated by rigorous observation. Rather than a victory of religion over science, this will be a victory for both science and spirituality, opening the way to the deepest exploration of human nature and our capacity for realizing genuine happiness through knowledge of ourselves and our relation to the natural world as a whole.

Acknowledgements

It is with deep gratitude and joy that I like to recognize all those who have contributed to this work.

To start, I express my great respect and thankfulness to Alan Wallace and Clifford Saron, who have been the initiators of the Shamatha Project. Their dedication, efforts, and heart for the cause truly shepherded this project from a concept to a reality. Thank you for your inspiration and support in the process of the evolving of this book. In expressing my appreciation to the science team members, I like to specifically recognize Baljinder Sahdra for her contribution to the project's ambiance. My gratitude goes to all those who made this expedition possible, including organizers, scientists all over the world, and sponsors.

Crucial team members to the project, and for me, have been the research subjects: the co-guinea pigs with whom I shared and celebrated three remarkable months, through ups and downs. Thank you for your presence, for our holding field! In this, I like to specially name Jim Cahill, Allie Rudolph and Michael Dysart, buddies on the expedition.

I am indebted to Professor Peter Harvey, who has been a very stimulating mentor in Buddhist Studies, teaching a Buddhism that clearly includes relating with present day concerns, "in the world and not of the world." His inspiration manifests in my description of possible societal applications of shamatha in today's world.

A word of thanks goes to the scientists and contemplatives who encouraged this writing with valuable comments at an early stage, at the Mind and Life Summer Research Institute 2008: specifically Joan Halifax Roshi and Richard Davidson.

In a deeply existential way, I bow to the wise women and men in all contemplative traditions on whose shoulders I am standing, in the privileged position of practicing in contemporary times with access to so many precious sources. I honor the teachers who guided me on the path: in Zen context Gesshin Prabhasa Dharma Roshi and Jiun Hogen Roshi; in Vipassana Venerable Jotika Hermsen and Dhammananda Bhikkhuni. In Tibetan Buddhist and Essential sense, I especially thank Lama Karta, Lama Palden Drolma, Rahob Tulku Thupten Kalsang Rinpoche, Daniel P. Brown, Gretchen Nelson, Susan Mickel, and Geshe Tenzin Wangyal Rinpoche, for their precious teachings. In expressing my gratitude to the Ridhwan Path I like to recognize Hameed Ali, Jessica Britt, Sherry Ruth Anderson, Eugene Cash, Nancee Sobonya, Eva Jansen, and the cherished members of our postal code group.

In direct connection to the present book, heartfelt gratitude goes to those who read parts of the text, gave their comments and shared their considerations. They include Marina Alers, Gretha Myoshin Aerts Sensei, Scott Virden Anderson, Ilse Bulhof, Jim Cahill, Sean Esbjörn-Hargens, Ingrid Foeken, Hiroko Ishikawa, Pamela Koevoets, Wolfgang Lukas, and Wouter van Waning. With this, a special thank you goes to Lama Palden Drolma, for much valued clarification, and to Sherry Anderson, for both sweet and empowering encouragement.

Teachings in psychiatry, psychotherapy, psychoanalysis and complementary approaches in body-mind work resound in this book. In this regard I would like to thank Niek Treurniet and Guusje Wolffensperger-Rübsaam, in appreciation for their open minds. Teachers arise in many forms and among these I like to recognize the patients-clients with whom I practiced and learned, and those who I've had the honor to guide on a meditation path. Thank you for continuing to touch my heart-mind.

There are some persons that I'm deeply grateful to for their specific form of attuned support, that I would like to recognize: Takeshi Gotoh, Marre Leijnse and Paul van Waning; also in this sense I bow to Ingrid Foeken, for being a dear longtime *kalyanamitra*, spiritual friend on the path. Thanks to my parents, relatives, friends – to all of you, whose interactions in some way nurture this book, in this wondrous interconnected world.

Introduction

> *What does it mean, "settling the mind in its natural state"? Who am I, who are we, what might be the natural state? The least thing I can do is to bring a number of fresh notebooks and pens for jotting down, reporting.* (Diary, September 2007)

In this book I present the story of my participation in the Shamatha Project, addressing Buddhism, concentration-calm practices, and meditation research. With diary fragments, dream log, and audio transcripts communicating my personal experiences I hope to give you, reader, a feel of how it is to be participating in such a project. I present the up-to-date research outcomes, that focus on the effects of the various practices on attention and emotion regulation, and on health. They include groundbreaking findings of effects up to the chromosome level.

Every chapter in this book includes a guided meditation. The book is structured in a way that it can provide you with various threads. You can read it, for instance, as an overview of the Shamatha Project, with a window into one participant's experiences with the project and the practices. Additionally, you may read it as an exploration into Buddhist studies, with a focus on psychological and scientific understanding of meditation. Most importantly, however: you can read it as a personal journey for yourself, combining it with taking up the meditation practices, with the guidance that is given throughout the book, including responses to questions that came up for us participants. So: you may immerse yourself into the practices and embark on your own nourishing journey, with guided meditations. In that way you will be cultivating your mental balance and well-being. This I can already give away: the science findings show wide-ranging benefits of the retreat experience for the participants. For myself, personally arising in flesh and blood body-mind out of the statistics database, I can only affirm this, in my felt sense experience!

How did I get to be a participant? Let me start with the beginning.

Once upon a time

In September 2006, in a far and desolate, yet inspiring corner of the Netherlands, a course was offered: "The sense and meaning of meditation: an interactive retreat with Alan Wallace." Located below sea-level at Naropa Institute in Cadzand, we participants shared Wallace's enthusiasm about the Shamatha Project that would combine three months of residential meditation retreat with state-of-the-art scientific research. Shamatha refers to concentration-calm meditation. Two shamatha retreats of three months' duration were going to take place in 2007. Some of my deep longings were coming together here: for quite a time, I've had the wish to participate in a meditation retreat of several months' duration, and combining this with research would add an extra fulfilling dimension to it. Inquiring about the possibility of participation, I received Alan Wallace's declining smile: oh, no, the waiting list is already too long …

Months later, I just thought: nothing wrong with being placed on a long waiting list, without much hope that anything would come from it. And then, at a certain moment I happened to be the person for the matching of the groups on demographic, meditational and psychological variables (including age, sex, education, and meditation experience) that was needed. Just a few weeks before the start of the Fall retreat I bought my flight ticket, and in September 2007 I happily found myself back on the shamatha cushion, hooked up to a monitor, 8000 feet up in the rugged Rocky Mountains.

In what follows, you read about my questions, wonders, confusions and understandings, in these three months of meditation and science, and daily "ordinary life," during this extraordinary period of time. Next to reflections from this period, I include diary musings from more recent times, showing how the project continued to ripple through my life. I will focus specifically on one form of the shamatha meditations that we practiced, "Settling the mind in its natural state." This one practice has interested and fascinated me specifically: as a personal interest, and connecting with my work-background in "minding the mind" and understanding the mind, with others, in psychotherapy and psychiatry. The name of the practice sounds intriguing to me, to the Western ear. I'm aware that I can have ideas about these notions, and I wonder about my culture and time-bound understandings. How do they relate to this Tibetan practice that originated many centuries ago, far away? What might it be, the natural state? This question will resonate and accompany us, sometimes more on the foreground, sometimes more in the background.

"The less dust, the more trust"

The less dust, the more trust: these words are heading the title of this book. They combine into one of the one-liners or little mantras that just came up during retreat, while I went for a walk in the mountains. It was a clear day, I was enjoying breathing this crispy Rocky air, and viewing the radiant, contrast-rich autumn colors. From the Shambhala Mountain Center, where we lived, many paths go up the foothills. I often walked up, in this magic valley scenery, through the aspen and then pine tree woods, then higher up where just rocks remained, with an immense sky above. Wind on my cheeks, and increasingly wide, fresh, clear scenery, that invited for freshness inside. To me, it feels like I still carry these mantra words with me, as a body-mind anchor for the whole retreat experience, including sense impressions, perceptions, musings, memories of feelings, people, interactions, some of them to be described.

In the mantra "The less dust, the more trust," the dust refers to the expression: "beings with little dust in their eyes," meaning beings who will be able to understand the Buddhist *Dhamma* teachings. The words refer to those who will be able to understand the truth of phenomena, including themselves. *Dhamma* is the Pali term, in the canonical language of Buddhism, *Dharma* the term in Sanskrit, the universal language of ancient Indian culture. The expression of "beings with little dust" figures in the *Ayacana Sutta*, "The Request." This *Sutta,* discourse of the Buddha, tells about the Buddha Shakyamuni, who after his awakening at first was hesitant to teach *Dhamma*, as very few beings might understand his teachings. Then the compassionate god Brahma Sahampati, aware of the deep suffering in the world and the need for this wisdom, convinced the Buddha to teach the *Dhamma* to beings in the world, with this plea: "Let the Blessed One teach the *Dhamma*! ... There are beings with little dust in their eyes who are falling away because they do not hear the *Dhamma*. There will be those who will understand the *Dhamma*." Luckily, as it is said, then the Buddha, out of compassion for beings, surveyed the world. "He saw beings with little dust in their eyes and those with much, those with keen faculties and those with dull, those with good attributes and those with bad, those easy to teach and those hard." [1] Here, we may all feel included. Any lesser particle of dust, every particle removed or dissolved will contribute to our seeing in more real and realistic ways, and acting more wisely.

Regarding trust: this certainly included my personal sense of trust, or lack of that. Interestingly, most traditions in Buddhism see trustful confidence or faith as an important quality. Trust must be balanced by wisdom, and can be seen as an accompaniment of meditation. It begins, in my experience, with becoming aware of this unease, the agitation, the reactivities in our ways of addressing ourselves and others. And then, somehow trust in the Buddha, his teachings and the community of those who have followed and realized the teachings can become a starting point for embarking on a path, a quest. This may regard the Buddha, or other wise and compassionate teachers. If we didn't have some initial trust in the fact that there is a way out of suffering, and some seed of understanding the nature of suffering, would we ever begin to search a path out of suffering, with any hope of finding it?

On my way

A few words about my on-going journey on the path, about what went before, are appropriate here. During my whole life I've been intrigued and fascinated by the mysteries of the mind. Living life has been the laboratory for exploration, reflection and deepening awareness. Trust and dust: there have certainly been times that I had little trust in myself and my experiences. Getting to trust ourselves and our experiences includes that we get to see the many ways that something in us may lead to deceiving ourselves, in distorting perceptions, sensations and emotions. For periods in my life this has been the case for me, generating and maintaining a dusty atmosphere. Both the sense of intrigue, and the suffering brought about by these distortions led me into psychotherapy as a patient. Enriched with inspiring therapists and teachers and with some tools for living a more truthful life, still intrigued and fascinated, I trained for being able to help others. In psychotherapy and psychoanalysis practice, I met others like myself, with their sufferings, unease, stress and reactivities, seeing how we all share in living our lives with various degrees of ignorance and distortions. I've been engaged in the psychotherapy field, with persons who, next to having mental and emotional problems, also showed immense potential. I had the privilege to work with adults and children, and for a time specifically with people with a history of migration, loss and homelessness. Together with these involvements, there were many years of meditation and spiritual practice, in various traditions. These, I feel, have been crucially important in my development and unfolding and being more naturally present in life, including work relations. With the mindful psychotherapy and meditation flows through my life and veins, my focus of interest and research have increasingly included persons' surprising developmental capacities and resilience in the face of hardship. There may be serious mental problems. However, there's always this unique, impermanent balance that also includes an impulse for development, maturation and evolution, with naturally manifesting greater compassion, including for oneself. These wholesome capacities and abilities can be cultivated, trained and strengthened. This we can experience ourselves, we can see this before our eyes; but we do not read about it enough in most textbooks. These capacities and abilities are much needed in this world that has witnessed and is witnessing so much pain, self-limitation and suffering. Buddhism has graphically sketched the various domains. On the one hand, there are these domains of suffering that are inevitably part of life (including ageing, sickness, death). On the other hand, there are the domains that are much man-made, woman-made, "reactive mind-made." Unwholesome traits that have been "made," conditioned, can to a large extent be de-conditioned. Wholesome traits can be trained: this is one of the positions that, I hope, can be strengthened with this book. As it is said: train your mind, change your brain. Neuroplasticity and the potential for transformation are "on" and in action all the time. Neuroplasticity can be invited for training resilience in pain, and for cultivating loving kindness, genuine happiness and trust. Training and cultivating attentional trust turns out to go in synergy with allowing dust to dissolve.

What you will find in this book

Let me elaborate a bit more on what weaves together in this writing.

First, I draw on my own first person subjective impressions as a participant in the Shamatha Project and on later understandings, often in diary form. The scientific aspects of the Project and Buddhist meditation that are addressed include the objective recent research outcomes. They include findings in follow-up measurements, and also plans for new follow-up data collection yet to come.

Second, as to Buddhist background, I turn to traditional texts by Tibetan masters including Lerab Lingpa and Düdjom Lingpa, to commentarial literature, and to verbal exchanges with teachers. Next to honoring my Western upbringing and education in some psychiatry and psychotherapy musings, I include insights from the Integral approach, to provide a map for organizing the various perspectives taken in this exploration.

Third and most importantly is my wish that you feel invited to join on this expedition in a felt-sense way. The meditation instructions by Alan Wallace guided us participants while we were in retreat. You may read them, put the book aside and practice "with us," in a timeless dimension. While sitting in your chair or on your cushion, you may join me in "going on an intensive and enriching meditation retreat," and in continuing the journey that this retreat inspired me to; and, with that, continue on your own path. So, in case you like to join: imagine yourself preparing for participating in a long-term meditation plus science retreat. How would you proceed in the practical sense, what to bring, what not to bring? And in the mental and emotional sense: what would you like to know, to ask, to read, what would it be that you might worry about? You might want to take in some background information. At this point you don't really need to do, as this book is going to give you some, in the process. Come in fresh, with "beginner's mind"! You may be aware of the turbulence in your mind from time to time. Would you like to know how it can be "tamed," as it is named?

In this personal report, I will most of the time follow the sequential way in which my understandings developed and unfolded. So, there I came in, carrying my Western background, into a center and group that was going to practice meditation in some sort of Tibetan Buddhist context. Sitting in a shrine room, with an altar, we were surrounded by colorful *thanka*s, silk paintings, depicting Buddhist deities. Right in front of where I sat, on a distance of a few meters, I looked at an impressive large one, with abundant gold embroidery. On the one hand, interestingly, we were immersed in the meditation perspective with instructions and context from the Tibetan tradition. Here the subjective experience aspect stood out. On the other hand, within the first days we were thrown into the language, agency, and "culture" of Western science. We were wearing EEG, Electro Encephalogram-caps on our head for brain research that was aiming at objective measurements.

For the diary sections, sprinkled throughout the book, most of the time I roughly and simply refer to the months: September, October and November 2007. These, globally, were the months of the fall retreat. For giving some indication of time frame for later diary sections (after retreat) I mention the year. Here's the larger diary fragment from which a few lines were presented at the start of this Introduction.

Shambhala Mountain Center, Rocky Mountains, Diary September 2007

> At this time my main research question can be named, very broadly: What will it be like for this participant to be in the project, when coming in with as open mind as possible. What does it mean, "settling the mind in its natural state"? Who am I, who are we, what might be the natural state? The least thing I can do is to bring a number of fresh notebooks and pens for jotting down, reporting. Anyhow, first step will be presence, awareness, and the aspiration to have an open look and describe phenomena as much as possible with "beginners mind."

Clearly, what I present in this book about the Shamatha Project are experiences of only one participant in the Project. As such they are not in any way representative, neither typical nor special: tens of first person stories and broader experiential reports can be written about the Project. As it stands, this is just one example, one story, by a grateful woman on the path.

Chapters overview

Following this Introduction, the initial chapters are directly connected with the Shamatha Project. The story continues into further explorations that I felt invited to by the project experience. Then there is a convergence in what I see as possible applications of the practices described, with a rounding up in some personal Conclusions.

Chapter 1: In this chapter, an outline of the Shamatha Project, initiated by B. Alan Wallace PhD, meditation teacher and Contemplative Director, and Clifford Saron PhD, Scientific Director, is presented. During the three-month retreat, we practiced shamatha attentional meditations from both Theravada and Mahayana traditions. Regarding the research, I provide a description of biomarkers, assessments, and tests. I report on the outcomes of the project focusing on the effects of the various practices in attention and emotion regulation, and on health. They include unique findings, among them linking meditation and positive psychological change with changes on the DNA level.

Chapters 2 and 3 offer a fuller presentation of the specific meditations involved in the project: attention practices with various aspects of tactile sensations of the breath as the object of mindfulness, "Settling the mind in its natural state," Awareness of awareness, the practices of the Four Qualities of the Heart (Loving Kindness, Compassion, Empathetic Joy and Equanimity), and *Tonglen*. While I present diary fragments and guided meditations in all chapters, these two chapters especially, together with Chapter 5, represent my direct experiential reports of being on this expedition. Included are verbatim transcriptions of guided meditations with Alan Wallace, with his instructions on selectivity and balance, mindfulness and introspection-monitoring. At the end of Chapter 2, I present some basic information on Buddhism, representing in this way – Buddhism, science – the main culture fields we were exposed to, at that time. For concluding Chapter 3, I present four working hypotheses, that guided my explorations regarding shamatha and Settling the mind.

Chapter 4: Here I describe shamatha practice in the Tibetan Buddhist context. In particular I explore how shamatha is the indispensable complement to the cultivation of contemplative insight, vipashyana. I address the nine stages of meditative concentration, the Elephant Path, and the notion of substrate consciousness, the relative ground of the psyche. Attention is given to various ways of understanding mindfulness and (meta) awareness.

Chapter 5: In this chapter descriptions of my personal experiences are presented, with first person phenomenological impressions of mind-operations and dynamics in the psyche. While regarding all practices that we did, the diaries especially address "Settling the mind in its natural state." The object of mindfulness in this practice is the space of the mind and what arises in it, with non-distraction and non-grasping: in the sense of no control, no preference, no identification. Grasping to a thought is released, while the thought is observed. In my descriptions I include diaries, audio, video and dream-log content as empirical material. The felt sense data include descriptions of inner turbulence while progressing with the practices, with surprises, ups and downs in the physical and psychological sense.

Chapter 6 offers meditation instructions and experiences regarding Settling the mind practice as presented by nineteenth century Dzogchen masters Lerab Lingpa and Düdjom Lingpa. Specifically, the turbulences that are bound to appear in some practice phases are addressed. A section from *The Vajra Essence* text by Düdjom Lingpa offers descriptions of turbulences and challenges such as intense pains and paranoia, anger and bliss. I've been intrigued by these descriptions of experiences, designated by Düdjom Lingpa as "signs of progress" on the path. I connect them with my personal experiences, and with descriptions of these dynamics by other teachers about their students. The turbulences as described in the text by Düdjom Lingpa, and as reported by myself and others seem to attest to universal dynamics in the psyche.

In the chapters that follow, I explore shamatha, including Settling the mind practice, in various kinds of larger contexts.

Chapter 7: In this chapter I explore the practice "Settling the mind in its natural state," a practice of "abiding, moving and awareness," in greater detail, in the context of being the shamatha part in Mahamudra and Dzogchen traditions. The less dust and the more trust, that shamatha brings in: together they make a foundation for these Essential insight approaches.

Chapter 8 addresses some psychological themes and explores the phenomenological descriptions, connected with supposedly universal dynamics (as presented in Chapter 6) in the contextual framework of both Buddhist and Western psychology. Here I explore the Buddhist notions of The Six Realms and The Four Maras, and some Western views as presented in the reference book the "Diagnostic Statistical Manual," and developmental stages in defensive maneuvers. The diary experiences connect in surprising ways with the broad range of defensive maneuvers and reactivities as described.

Chapter 9: This chapter addresses mental turbulences on the meditation path, including the themes of spiritual emergence and emergency. This is about "falling out of habitual conditionings," about the grasping self, and the potential for opening up. Attention is given to energy dynamics, with contraction and relaxation. I explore the ways in which "Settling the mind in its natural state" can be conceived of as a form of psychological therapy of self-healing. I elaborate on what has been a growing realization about a continuum, from psychological dysfunctioning to exceptional health, knowingness and insight; challenging the sense of "normalcy" of our collective habitual level of consciousness.

Chapter 10: Here some themes in contemplative neuroscience are taken up, including shifting baselines and the notion of "acquired secure attachment." The practice of "Settling the mind in its natural state" is placed in a research context of two basic types of meditation: Focused Attention and Open Monitoring. A model protocol for phenomenological meditation description in a neuroscience context is applied to the Settling the mind practice. Two pilot projects regarding shamatha and Settling the mind practice are described.

Chapter 11 includes my revisiting the four working hypotheses regarding shamatha and specifically Settling the mind practice that I presented in Chapter 3. I describe what I came to find and understand. In this chapter I offer some ideas on shamatha in the world, possible practices and applications for special groups of people, including persons with attention deficit problems, persons with psychiatric difficulties, and mental health professionals; and applications for those who guide meditators on the path (including going through turbulence), and as a preparation for dying. In this chapter I also present information about recent developments in the Shamatha Project research, including plans for a new round of follow-up measurements to come.

Chapter 12: in "On human flourishing: continuing the project," with presenting rounding up remarks and afterthoughts, the theme of dust and trust is revisited. Various meanings of Settling the mind in its natural state are addressed, as related to the vantage point from which one views. Here I come to four aspects that have touched me in a personal sense in this quest. One relates to the many ways in which I have experienced Settling the mind practice as a gift in my life. The second aspect regards the way that Settling the mind has facilitated for me experiencing a wide range of mental states, from coarse emotional, to subtle, to very subtle. Three: I'm aware that, with observing and experiencing various states of mind, I've trained traits of resilience. I have experienced the healing and therapeutic aspects of shamatha practice in general, and Settling the mind in particular, translating into character traits with greater openness, maturity and freedom. And Fourth: I've come to value more deeply how the findings in the many domains of investigation regarding meditation, including phenomenal experiential and their neural aspects come together; being correlates, that can neither be separated, nor reduced to each other.

Mindfulness and the mysteries of the mind

With our minds we get insights into the workings of our psyche. With our minds we can change our relationship with the mind, with the arisings in the mind, and our perception of reality. Thus we can change our relationship with the world, and open up to a deep sense of joy and well-being. With our mind we can rewire and transform our brain. Our increasing understanding of brain processes, perception and attention, behavioral and psychological dynamics, has greatly inspired our ability to fathom, prevent and remedy problematic health conditions.

Mindfulness, moment-to-moment attention, in this shamatha context mainly refers to non-forgetting what we are focusing on. Mindfulness goes in synergy with a meta-awareness in monitoring the practice. In various traditions different accents are placed with meanings of mindfulness. Celebrating the precious gift of mindfulness: in this book mindfulness will also be addressed in the context of vipashyana and of contemporary mindfulness based approaches.

In line with the multi-perspectival approach, this work is meant to offer a balance between three perspectives. There is the perspective of my personal experience: this is a subjective, first person perspective. There are the science and theory aspects: objective, third person perspectives. Next to that, I invite you to practice and explore the meditations for yourself as you read this book: this can be referred to as an intersubjective, second person perspective.

It has been a great joy for me, with a background in psychiatry-psychotherapy, in Buddhist meditation and Buddhist studies, with an interest in consciousness studies, and with a feeling of urgency for more inclusive and integral understanding, to have the opportunity to explore these themes. The start of it has been in the context of writing a thesis for Buddhist Studies about the Shamatha Project, with special attention for the practice of Settling the mind in its natural state. While I did this with great pleasure, there was a deadline, and a maximum word-count, both inducing some effort. Curiously, after finishing there was just the feeling that a book wanted to be written through me, which took some sort of an effortless start. There was some understanding, generating questions, inviting for new, more inclusive understandings, raising new questions, experiences, curiosity, feeding the next step and the next, all along. Then, letting it all percolate, integrate, and then release. May the writing and reading of these chapters make a little contribution to less dust and a clearer view.

Chapter 1

The Project, the research, and some outcomes

> *Our scientists are very committed people, handling their subjects with great care and respect. Still, it's an interesting experience to be aware of a very functional aspect in the relationship, in which we participants act, and can be seen as the providers of data ...* (Diary, 2007, 1.4)

Inviting people to act as research subjects, in a secluded barren setting, without distraction, and without external contacts, for three months of meditation and science ... who will apply for such an endeavor? This was one of many things the initiators of the project wondered about. Well, around hundred and fifty eager meditators applied, including me.

In this chapter I introduce the Shamatha Project that combines intensive meditation from an ancient tradition, with state-of-the-art neuroscience research, to study meditators who live an active life in society. A long period of incubation and preparation has preceded the project's manifestation in the two three-month residential retreats, in spring and fall 2007.

In the pages that follow I describe the way I was informed about the project and the way the retreats were set up. Then I give an overview of the scientific research program. I elaborate on the felt sense of being a scientific guinea pig, and present outcomes of the research, up till now.

1.1 The Shamatha Project, an outline

"Shamatha," in this context, refers to: meditative quiescence, calm abiding, tranquility meditation, or concentration-calm meditation. Different terms for shamatha are shared by all Buddhist traditions. There is a term like *samatha,* or *shamatha* in both Pali and Sanskrit.

The Tibetan term is: *zhi gnas*.[1] The sound is "zhiné." Shamatha is about directing the attention to an object of attention, with continuity and completeness, so that the attention "stays" on the object. This to me feels like a bit broader meaning than concentration and calm – still, as these are the designations usually applied, I'll keep to that naming. Shamatha is considered an indispensable foundation for the cultivation of contemplative insight – *vipashyana*, vipassana, in Sanskrit and Pali – and for the Tibetan Buddhist Essential practices of *Mahamudra* (Great Seal) and *Dzogchen* (Great Perfection). These will get more attention in later chapters. Training in meditative quiescence cultivates relaxation, attentional stability and vividness of perception. As the techniques can be performed in quite structured ways, and as they do not demand allegiance to any religious or philosophical belief system, they lend themselves well for scientific investigation, and for participation by a wide range of potential subjects, especially in the secular West. Alan Wallace, our teacher, considers shamatha meditation a potential bridge between contemplative and scientific means of exploring the mind.

Regarding meditation: many different descriptions are used, referring to a wide variety of practices. They range from relaxation exercises, and exercises designed to improve attentional function (like in concentration-calm), to aims like the cultivation of greater well-being, equanimity, and compassionate behavior, on to awakening for the sake of all beings. In a research setting meditations may be described as complex emotional and attentional regulatory strategies. In this book, emphasis is on concentration-calm meditation practices and practices for cultivating qualities of the heart, like loving kindness and compassion. They are presented in the larger context of practices aiming at freeing ourselves from dust and veils, aiming at liberation.

The Project enterprise

The Shamatha Project, so I learnt, is a joint enterprise of the Santa Barbara Institute for Consciousness Studies, University of

California-Davis, and Shambhala Mountain Center, all in the US. Alan Wallace PhD, Contemplative Director, and Clifford Saron PhD, Scientific Director, the initiators in this on-going project, started preparing for the project around 2002. The seeds for the project had been waiting since long before that for the right moment to germinate.

The full title of the project is: "The Shamatha Project: A Longitudinal, Randomized Waitlist Control Study of Cognitive, Emotional, and Neural Effects of Intensive Meditation Training." An inviting advertisement text, meant to attract potential participants, included the sense that the participant would become part of (contemplative) history: "Meditate to Advance Science – be part of this groundbreaking neuroscience research project exploring the relationship between meditation and well-being." The readers, including me, were informed that Wallace would guide participants in various forms of shamatha practice drawn from the Theravada and Mahayana Buddhist traditions. The announcement presented comments of His Holiness the Dalai Lama: "I believe this research project has the potential to be of significant benefit for advancing scientific understanding of the effects of meditation on attention and emotional regulation."

The reality is that early on in my medical studies I have had to work with actual guinea pigs. They had no choice, and I still feel pain in my heart thinking back to those days. As it is said in the Buddhist tradition: they all may have been my mothers and fathers. I had a choice now, and even if I didn't know what exactly that would mean, I didn't hesitate to function as a scientific guinea pig myself.

As I knew, and got to learn more about: B. Alan Wallace PhD lived as a Tibetan Buddhist monk for fourteen years and was ordained by His Holiness the Dalai Lama in the Gelug tradition (one of the main schools in Tibetan Buddhism). He has a broad and varied background: he earned his undergraduate degree in physics and the philosophy of science and his doctorate in religious studies.

Currently he is the Director of the Santa Barbara Institute for Consciousness Studies. He writes books and articles and leads meditation retreats worldwide, like the one that I initially participated in, in 2006 in the Netherlands, before I joined the Shamatha Project.

Clifford Saron PhD is a neuroscientist at the University of California, Davis, Center for Mind and Brain. He has had a longstanding interest in the effects of contemplative practice on physiology and behavior. Next to his work as scientific principal investigator in the Shamatha Project, his other major research area is a focus on sensory processing and multisensory integration in children with autism spectrum disorders. In new research in collaboration with other groups, he is combining these strands of work, exploring how mindfulness based interventions can ease the chronic stress of mothers of children with these disorders in ways that may be beneficial for the whole family system and contribute to a lessening of difficulties for the affected children.

Since around 2002, Wallace and Saron, with a team of over 30 investigators and consulting scientists, have been involved in planning the Shamatha Project in a very broad set up that would generate a huge amount of data. Saron, in retrospect, describes the Project as unique in the aspect of including measuring telomerase in the meditators. In that way, research on the DNA level is part of the project. Telomerase is the enzyme that repairs telomeres. Telomeres are the very tips of our chromosomes, and a marker of stress and cell aging. In addition to the telomerase research, the project has been unique, Saron feels, because it was "science driven." As he shares with health and science journalist Thea Singer: "A research team has never set up a three-month retreat before ... The way this has always been done is that the retreat is set up by a meditation center and the researchers show up like filmmakers do and film an event. They have nothing to do with who's in that retreat or with the structure of it. We did something completely unique, which was work with Alan to create essentially an admissions committee, the advertising material, and the retreat logistics. He was in charge of what was taught and the schedule of practice. We were in charge of the testing." [2] The project evolved into this form of two retreats, with state-of-the-art scientific measures in a randomized wait-list controlled study. The meaning of this is that during the first retreat, the spring retreat, the research team studied the trainees, while at the same time measurements were done on the members of the control group (not meditating more than their usual daily practice, during this period). These persons then became full-time participants in the second retreat, the fall retreat. Practically, this meant that participants of the fall retreat were flown in during the spring retreat, coming over for just a few days of acclimatization and measurements. This set up made it possible for the researchers to use the control group members, first, for comparing the results with those of the retreat participants, meditating full-time. Second, the control group members' results when they were not on retreat could be compared with their own results when later they were on their own retreat. With some others, I participated in the fall retreat without having had the opportunity to do control measurements during the spring retreat.

For the interested person, it is possible to check out online the set-up of the project and all scientific articles' texts that have been published so far, and that will be published in the future. [3] A summary-overview of the research outcomes, described in the articles that came out so far, is presented in 1.5. This includes publications by groups of researchers with Katherine MacLean, Tonya Jacobs, Baljinder Sahdra and Manish Saggar as first authors.

1.2 Inspirations, preliminaries

One may say that the Shamatha Project, in its embryonic form, started more than twenty years ago, in Alan Wallace's musings and aspirations while being a monk. As he told us, he was contemplating, then, how meaningful it could be to set up a Shamatha retreat of one year. Having himself thoroughly been trained in shamatha practices, and having experienced the great benefits, he would like to bring more Westerners in contact with this approach. In 1988 Wallace served as the assistant and translator for Gen Lamrimpa, one of his Buddhist teachers, who gave a shamatha retreat in Washington state, US. Twelve participants completed the one-year retreat. In those days, as Wallace reflected by hindsight, the time wasn't ripe yet for combining such a retreat with scientific monitoring and research.

In 1992 a group of enthusiastic scientists, with the support of the Dalai Lama and the Mind and Life Institute (an organization devoted to establishing a working collaboration and research partnership between modern science and Buddhism), took measurements with highly trained contemplatives up on the Himalayan slopes around Dharamsala, India. This has been quite a step in the scientific sense: conversing and studying with the contemplatives in their own environments, with instruments from the mobile laboratories carried in the scientists' rucksacks. Alan Wallace and Clifford Saron were the prime organizers, embarking on this expedition with Jose Cabezón, Francesco Varela (cofounder of the Mind and Life Institute), Richard Davidson, inspirer of much meditation research to come, and others. Since the late eighties, with the help of refined neurophysiological and neuro-imaging techniques, it has become clear that, contrary to earlier assumptions, the brain can adapt, heal and renew itself to a high degree. The range of neuroplasticity, the ability of the brain to reorganize itself by forming new neural connections, is astounding. Training in any field, such as music or sports, can significantly alter connections among neurons and modify the brain systems devoted to particular tasks. The baseline is that repetitive practice, whatever you are doing, including mind practice, changes the brain. The brain optimizes itself to the milieu it is in, to its "environment." Recent studies are adding meditation to the list of training that can potentially change the brain in beneficial ways. The effects of meditation on coping with stress, on emotional stability, and of meditation for therapeutic applications are now legitimate topics for rigorous academic research.

Many studies have mainly addressed effects after a relatively short term of practice. Other projects, increasingly, are with people who have had many years of intense meditation experience, such as monks or nuns. [4] The Shamatha Project adds to a growing number of studies on "ordinary" people who live active non-monastic lives in society. The Project is unique in that it uses an exceptionally broad range of measures regarding attention, cognitive performance, emotion regulation, health and well-being.

Taking risks

As Wallace remembers, all involved, including the scientists, took risks: what will come from this? One of Wallace's concerns was: who wants to watch their breath for three months? Maybe just ten people will show up, and half of them may not have the appropriate background. Indeed, then what? It turned out to be no problem. For the one hundred and forty-two persons that applied in a self-selective way, selection criteria included that they had attended at least 3 prior retreats of 5 days or longer and that one be a shamatha retreat with Alan Wallace. A ten-page application form with information about meditation experience and current daily practice, psychological health, physical health, personal goals and motivation, had to meet the requirements. To name one concern in the selection procedure: can the potential participant stand to be mostly in silence and solitude, to be self-sufficient and disciplined, to tolerate minimal contact with the outside world? After selection on the basis of paperwork and cognitive tasks, a psychologist took all aspiring participants through psychological testing by telephone. If required, as in my case, intercontinental connection was made. In this way, out of the group of applicants, sixty were selected by the project staff, for two groups that had thirty participants. Ages ranged from 22 to 69. Our group of participants in the second retreat finally consisted of sixteen women and fourteen men.

The image that Andy Fraser sketches of the set-up of the Project, in an article about the Shamatha Project in 2011, brings up a smile. Fraser imagines as follows: "It could be a scene from a science fiction movie. A group of scientists converge on a remote homestead in America's Rocky Mountains and, in a frenzy of activity, assemble two state-of-the-art laboratories in the basement of a quiet residential building. They install video cameras, instruments for measuring brain activity, heart rate and respiration, and a "blood lab" where samples can be taken for biochemical analysis." Yes, I remember my own first impressions, and for a person less prepared it must really have looked quite extraordinary. Fraser wonders: "Perhaps they are there to investigate reports of alien activity? Has a radioactive meteorite unleashed a horde of zombies on this previously peaceful community? Or maybe it's just the start of the Shamatha Project, the most extensive long-term study to date on the effects of meditation on the mind and body." [5]

1.3 Meditation expedition: overview, daily life

Wallace speaks of an *expedition* rather than of a retreat: there's nothing about re-treating, in this project, this is rather about an advance, it is about participating in an exploration into meditation and science that will bear fruit in daily life! It is not turning away from your life but really engaging in it. Shamatha training cultivates relaxation, attentional stability and vividness of perception: abilities that have their value under any and all circumstances.

During this Shamatha expedition, we practiced natural, optimal balance in four "spaces," or domains of perception, experiencing and consciousness: body, mind, awareness and "heart." Here is a very brief overview.

All practices done during the project retreat comprised training in directing the attention, in these domains:

1. Mindfulness of breathing, with as a focus: the tactile sensations of the breath in the whole body, the abdomen, and at the nostrils,

2. Observing what takes place in the mind, with focus: the space of the mind, and what arises in it. This is the practice "Settling the mind in its natural state," and
3. "Shamatha without a sign": Awareness of awareness, where all information from the five physical senses and the mind are left aside. There is not a sign, in the sense of focus, there is just awareness.

   These practices can be summarized, in Wallace's phrasing, as: soothing the body, settling the mind, and illuminating awareness.
4. The "Four Qualities of the Heart," also called the "Four Immeasurables": Loving-Kindness, Compassion, Empathetic Joy and Equanimity. These were complemented with *Tonglen* – the Mahayana practice of giving and receiving. Together they formed a heartful complement to the named attention practices.

Breath, whole body practice – guided meditation

Let's start right away with a guided meditation. The focus in this meditation regards the tactile sensations of the breath in the body. It's not important here whether you have ever practiced before, just engage in this practice to the best of your ability. What, for you, is the felt sense of breathing in, breathing out, and the turning-points between these, in your body? Here is a meditation of the "whole body breath" approach with guidance by Alan Wallace. Take a posture of ease and relaxation. More about posture and explanation of some terms that are used will follow in Chapter 2. This is a transcript of a guided meditation that Wallace gave us in the beginning of September 2007, one of the first that I heard on retreat. This guidance transcript and the transcripts to follow are all Wallace's instructions, unless mentioned otherwise.[6]

In this practice the *object of mindfulness* is the space of the body and whatever arises within it. How lovely it is to simply set the body at ease, surrendering the muscles of your shoulders to gravity, releasing all the tension in the face, especially around the eyes; and the eyes themselves, let them be soft and relaxed. When you are at ease, your body is comfortable. It can be still, apart from the natural movement of the breath.

Now to balance the deep relaxation of the body we sit in a posture of vigilance, with the spine straight, the sternum raised, sitting in attention. Let your awareness permeate the whole field of the body as you settle your respiration in its natural rhythm. This practice is a wonderful, wonderful respite. If at any time for any reason, whether emotional, or maybe digestive problems, or you didn't sleep very well, or just getting frustrated that you're not making enough progress ... In all of the above cases, doing whole body meditation is OK. Every *out-breath*, an opportunity for relaxing and releasing more and more deeply. Every out-breath, gently release any involuntary thoughts that have arisen, as if the out-breath is a gentle gust of breeze blowing away leaves. Just let them go. Continue letting go of the out-breath all the way through the end, releasing and releasing until you find the *in-breath* flowing in of its own accord, whether it's deep or shallow, fast or slow. Just let it be. This breath, that we can so easily regulate and alter by merely preferring, release it. Let the body breathe without intervention, without influence by your will, desires or expectation. Let your awareness be diffuse, permeating the field of tactile sensations like a gas filling a room. Quietly, non-reactively be aware of whatever tactile sensations arise within this field.

Observe whatever sensations arise without distraction, letting your awareness rest in the present moment, here and now. With each out-breath releasing all thoughts of the past and the future, all cogitations about the present. And observe these events within this tactile field to the best of your ability without grasping, without superimposing upon them conceptual images, without superimposing any notions of "I" or "mine." Simply observe them without preference, allowing them simply to arise in the space of the body, this field of tactile sensations.

In addition to the faculty of *mindfulness* we also refine and utilize the faculty of *introspection*, monitoring the meditative process, the quality of attention. If at any time you note with introspection that you're falling into dullness or laxity, when the clarity of mind is fading out, then freshly arouse your attention and pay closer attention. And with introspection, at any time if you know that you've been caught up in thoughts, relax more deeply, especially with the out-breath, and release both the thought and grasping onto it. And let your awareness descend once again into this field of tactile sensations ...

Let's continue practicing in silence ... Let's bring the session to a close.

Take your time for being aware of the felt-sense of your body-mind, while rounding up the session.

Setting, people

Most participants in our fall-retreat group came from the US, a few were from Mexico, one from Canada, one from the UK, one from Thailand, and me from the Netherlands. People were invited to arrive at a specific date, trickling in, in order for the scientists to do their research. It was stimulated, during the first few weeks, that we got to know each other a bit during mealtimes. In that way we could be a supportive group when diving into silence.

We were living, meditating, and being studied, in the Shambhala Mountain Center at Red Feather Lakes, north of Boulder. It is an isolated retreat facility on 600 acres of grounds in the Colorado Rockies. No public transport available around here: when I arrived I was collected from the village by a friendly community member in a pick-up car, with the last part of the journey going over unpaved dirt road. The center is situated in a mountain valley of pine and aspen forests, with alpine meadows, at an elevation of eight thousand feet.

The name Shambhala refers to the legend of Shambhala, teaching us that the basis of enlightened society is the understanding that human beings inherently possess wisdom, compassion and goodness. These qualities have been nurtured through contemplative disciplines in many cultures throughout history. So, this was a good place and atmosphere to be in, for us, doing practices that were meant to directly support our engagement in life in beneficial ways. The center has been founded by the late Chögyam Trungpa Rinpoche. Now his son Sakyong Mipham Rinpoche is the main teacher. The climate here is generally dry in summer, with abundant sunlight at this high elevation. At the same time, temperature may drop below freezing at any time of the year. So it did. While at the beginning of September we were meditating in burning sun, throwing off our clothes to (almost) topless, we had to suddenly shift later that month to many thermal layers. End of September: snowstorms and bitter cold. November showed up even more intense.

Diary, November 2007

> We live here with bears and with deer and chipmunks on stroking nearness, in a place where fox, mountain lion and coyote reside. In November we measure a temperature of minus 20 C, which is minus 4 degrees F. Curtains of icicles dangle down from the roof, shining in bright winter-sun. The light is so intensely clear, I need sunglasses, sometimes also inside the building. When walking, watch out for slippery ice-tracks …

Regarding our living quarters: our building had a roped exterior zone of silence, with signs saying "Shamatha Project, Silent Practice Area, No Public Access." It had been assigned specifically to the Shamatha Project, and was situated relatively separate from the rest of the Shambhala Mountain Center compound. We participants were requested not to socialize with other resident meditators, and to keep to our grounds. We needed to have permission for some sparing external telephone contacts and email. As little distraction as possible ... It reminded me of the play *Huis Clos,* closed doors, no exit, to say it with the French existentialist philosopher Jean-Paul Sartre. No escape from the task posed!

The two-storied building where we resided had an overhanging roof and verandas alongside. Our common meetings were in the shrine room, a lovely place with windows on three sides, bathed in light. It had a wooden floor, with an open corridor in the middle and on both sides a block of blue meditation mats and cushions, four or five, with four rows; some people sitting on chairs. In the course of the weeks, all of us increasingly made ourselves at home on our square meter, bringing in our own cushion, bench, scarf, cardigan, and water bottle within reach. In the front was the altar, a table in the characteristic Tibetan red color, with the seven bowls for water offerings, a bronze Buddha statue, flowers, candles and incense. The large gold-embroidered thanka with intricate texture showed the Buddha, with hands in the *Bhumisparsha mudra*: this is the hand gesture calling upon the earth to witness Shakyamuni Buddha's enlightenment. The seated figure's right hand is reaching toward the ground, with palm inward, and the left hand lies on the lap, with palm upward. For three months this has been my precious view. On just a few meters' distance on the same side of the room was Alan Wallace's sitting spot. Through the large windows I saw rough rocks and hard grasses nearby, far off mountains, with a bunch of aspen trees in the middle. An inspiring sight: viewing the colors changing, from full summer green, to bright yellow, sparkling gold, with soft pink, bleeding red ...

In the basement of the building we had two state-of-the-art psychophysiology field labs, a gym, and a room used for daily yoga sessions, led by one of the participants. Another participant in the project, who was officially appointed as our psychotherapist, could be consulted by those in need thereof. Most of us had a private room, in which we spent many hours a day in meditation. No other people were allowed in the building. There was, as much as possible, silence all over.

The large room where we had our meals was on the ground floor, bordering the shrine room. While for breakfast we had our foods stored in two refrigerators, warm meals were brought by pick-up car from the general kitchen. A memory: on the day I arrived it struck me that one of the refrigerators looked quite scarred, with striking unevennesses in its white metal walls on both sides. Wondering about that, I heard that the scars are witness to recent bear claws. Two days before I arrived, a mother bear and child had entered the building at early morning time (the door not being well locked), looking for food, smelling their way, and just throwing down the refrigerator. This happens, in autumn. One of our participants described how he woke up from hearing strange noises, was shocked when realizing what was going on. Later, seeing this fridge torn down, with all the food spilled over the floor ...

Indeed, there were bears around, I've seen a small one slowly walking over our veranda, in the evening. On various places on the grounds, information was given about "Living with wildlife: in bear country." For some this may not be so special, but for a girl from the Dutch flatland it sounds very impressive. So, I learned, for instance: never ever intentionally feed a bear ... store garbage outside in bear-proof containers (which might also be advisable inside). Also, the instructions taught me: "If you encounter a bear: 1. Back away slowly while facing a bear, 2. Speak softly to the bear and try not to show fear, 3. Fight back if a bear attacks you. Use rocks, sticks, binoculars or any object that may be available." Followed, with some empathy, by: "Remember, if a bear stands upright or moves closer to you, it may be trying to detect smells in the air. This isn't a sign of aggression. Once it identifies you, it may leave the area or try to intimidate you by charging to within a few feet before it withdraws." I was prepared.

Finding ourselves at home

These are some more diary lines capturing the beginning period of when I arrived and began to participate in the Project, September 2007:

> Many of us have given our home, the room that we sleep, sit, live in, a bit of a familiar flavor.
> Making myself at home ... The timeless text on my door is by Nyoshul Khenpo Rinpoche:
> "Rest in Natural Great Peace,
> This exhausted mind,
> Beaten helplessly by karma and neurotic thought
> Like the relentless fury of pounding waves
> In the infinite ocean of samsara."

It is about Karma, cause and effect, balance, and about the samsaric cycle of existence. Nyoshul Khenpo Rinpoche was an accomplished Dzogchen master, one of many Tibetan teachers who had to escape Chinese oppression in 1959. [7] "Rest in natural Great Peace" feels like a meditative invitation. After the long trip from Europe and all the excitement of starting the retreat I was eager to begin letting my exhausted mind rest in "Natural Great Peace" and to start practicing Settling the mind in its natural state. That was why I had flown 5000 miles.

Program

Meditation was the main course in each daily menu. We had two daily sessions with Alan Wallace and the complete group in the meditation hall. This meant: half an hour at 9 am for one of the Four Qualities of the Heart and *Tonglen*; and at 5 pm half an hour for one of the shamatha attention meditations. This was followed by one hour of questions and answers, sharing, and *Dharma* talk. I loved the contrast to silence that these lively and inspiring exchanges provided. In the first few weeks Wallace gave guided meditations, like in a "cineac system": every one or two days another one of the meditations was practiced. After a complete round, a new round followed, with more refinement, and different aspects. Later the practice was more in silence. All of us participants had a weekly fifteen minute interview with Wallace. He could also be approached in discussion sessions, and by way of a note on the clipboard.

Most of us meditated in our room, some shared a few sessions a day in the meditation hall. The average number of hours spent in sitting meditation turned out to vary from six to ten hours a day. As participants we took care of all the washing up and cleaning of the building, in rotating work assignments. Washing up was a general rota job. We got accustomed to silently putting on our aprons and rubber gloves, knowing our place in the oiled machinery. Next to that, I did a lot of vacuum cleaning along the endless corridors, on three floors. Also there was the rug-cleaning outside of the hallways where every day a huge mass of sand, stones, snow, and mud came in with our walking boots.

Going for a walk on the Shambhala grounds, through valley and mountains, was a stimulating activity and many of us did so, every day. It was permitted to share a walk with a shamatha buddy, and exchange about one's experiences. Some of us went for meditation to the Great Stupa, a sacred landmark on the grounds, said to be the largest representation of sacred Buddhist architecture in North America. It has three accessible floors, and is filled with original artwork. A few times we also meditated there with the complete group. Another every-day occupation was completing the seven-page Daily Experience Questionnaire for the research. In this questionnaire, specifically designed for this retreat, every evening we logged forms and duration of meditation, health, moods and feelings, important experiences and insights, and (next morning) dreams. This was quite a task, a daily awareness practice in itself.

Silence

While we were in silence, if a situation necessitated it, there could be some skillful speech, functional talking, in a way that this didn't reach others who were not involved. We had some community guidelines and we could make suggestions for additional ones. Guidelines were regularly updated, by way of exchanging written messages. The aim: a way of living smoothly together, in our silent Sangha.

Even in a radically simplified life (or rather because of it!), some issues may arouse quite strong emotions. Here is an example of a concluding message by the guideline-overviewer. It may read like a dry, factual outcome; however, it evolved only after heated exchanges, on paper and even in a special meeting: "After considerable discussion amongst those who use the shrine room regularly, it was decided that there would not be a fixed schedule for the optional sits. People will instead try to be as quiet as possible as they come and go." This also brought up a guideline for steps in possibly needed conflict resolution. Relating to this I find some remarks in my diary:

Diary, September 2007

> Impressive, the group phenomena that show up, in our group of people that hardly know each other while starting to live very closely and intimately together in silence ... It feels like our personal and "at home group" habits and attachments do shine through! As some of us here have a background in Zen, vipassana, Tibetan, Vedanta and other meditation tastes, this question about entering or not entering the shrine room has been quite an issue. I remember from the Zen groups where I have practiced: meditation session is sort of "holy time space." When not arriving in time, then to enter the Zendo a few minutes later is utterly not done! So indeed, I've been standing before a closed door. In vipassana groups, this could be done (with varying amounts of noise, and inner emotional response, as "grist for the mill" for noting and naming). The discord here: exemplary of our attachments! Good opportunities for practice of the Four Qualities of the Heart. It feels like the emotions may also be seen as connected with, for all of us, the uncertainties and loss of a

familiar hold, in the beginning phase of this expedition.

The last remark was certainly true for me, as I didn't know any other participant when I arrived. Next to this example of initial disagreement, right from the start there was also a striking movement of appreciation integration: many of us felt moved to offer our talents in some way to the community. A microcosm of different abilities was put in, as much as possible in silence. To give some examples: two persons offered yoga classes, one gave special back-problem mindfulness training. A dentist offered his services for people in need, as did hair-cutters. One participant taught meditative weight-lift classes in the gym room. An aspiring astronomer gave some practical celestial teachings, and an amateur photographer created a 600-picture photo-report of the expedition.

A warm feeling comes up in my heart when I write these lines. And looking back, I've felt so touched with the special experience of getting to know people in such a remarkable non-ordinary, non-socializing way, in silence. With the body language and the felt sense of a field of presence: what a miracle how in this field, a group culture and some deep personal bonds can develop and flourish!

1.4 Science: a brief overview of biomarkers, assessments and tests

Attention, in the sense of directed, sustained and domain specific attention has been researched in academic centers; but for a long time not so much has been known about the degree to which attention can be developed and trained. [8] The researchers in the Shamatha Project posed questions about the effects of meditation including these: what measurable changes in attentional ability occur as a function of intensive meditation training? What are the neural correlates of these changes and the range of their consequences? They wondered, about attention in relation to emotion regulation: is it true, as Buddhist contemplatives claim, that improvements in the voluntary control of attention and associated improvements in attention systems in the brain make it easier to recognize and overcome negative emotions, maintain resilience in the face of stress, and improve relationships with other people? And: do the changes persist after the meditation trainees return from the retreat experience to the cacophony of everyday life in a modern society? Next to what happens while one is meditating, there was a great interest in what happens because one has dedicated time to meditating. This refers to what one does in a different way, later, in daily life back-home, because one has meditated.

The scientists hypothesized that these three months of shamatha training, combined with cultivation of the Four Qualities of the Heart, will result in improved attentional performance (vigilance, selectivity in focused attention, and meta-cognitive control), as well as greater compassion and security, greater ability to down-regulate negative emotions, and apply prosocial values and motives.

The most comprehensive study to date

During both three-month periods, all participants were "measured" during two days each, at the beginning, then two days each halfway through the project, and then two days each at the end of the retreat. This regarded both meditators and controls. I remember that our science team of six members worked very, very long days during these periods.

Scientific measures, as summarized in professional language, have included established paradigms in cognitive and affective neuroscience, stress and affiliation-related biomarkers, Electro Encephalogram (EEG), physiological markers of autonomic nervous system activity, facial expressions of emotion, daily journaling, self-report and structured interviews.

It sounds like a big range indeed, the way it is stated in a News Release from the Association for Psychological Science in the US: "It's the most comprehensive study of intensive meditation to date, using methods drawn from fields as diverse as molecular biology, neuroscience and anthropology." [9] On the one hand, part of the measures used have been from well-known mainstream tests, to make comparison possible. On the other hand, no tests exist that address the attention in its refinement as might be measured in this project. So, as we participants were told, new, innovative tests had also been devised.

We paid for participation in the meditation retreat, for room and board (around 5,300 dollars). Next to that, we participants were compensated for the testing sessions for the studies (20 dollars per hour). I felt impressed by what we had to read and agree to: there's a lot that a human guinea pig has to learn and sign about her and his rights in a third person approach, as laid out in the "Experimental Subject's Bill of Rights – Social and Behavioral Studies." We were carefully informed about procedures, and seen as autonomous decision makers … Here are two fragments from the many-page document that I found interesting:

> *"What risks can I expect from being in this study?"*
> There are no physical risks to you from the experimental procedures. You may experience slight discomfort or a cool sensation when we apply special gel to your scalp and several locations on your face when preparing to record your brain waves …
>
> You may get somewhat fatigued when performing the attention tasks. You may experience a short, unpleasant emotional reaction in response to some of the emotional films or pictures you watch. During the meditation retreat you may experience emotional ups and downs and recall experiences from your life that you have not remembered for a long time …
>
> *"Are there benefits to taking part in this study?"*
>
> It is possible that you will not benefit directly by participating in this study.

The last statement felt like a really careful, anticipatory consolation, in the event that …

First, second and third person aspects

An interesting dimension of this project was the combination of third person measures (referring to research looking at objective measures) with first person research (the way an individual, I, subjectively experience things). The second person perspective (the intersubjective view of the participant by someone else) comes to us from those with whom we live; it is to be included in planned follow-up interviews with individuals close to us participants. While the video interviews were investigated according to a refined coding scheme, second person perspective impressions from the video interviewer and from Alan Wallace were not formalized in the research.

In first person research one is concerned to understand perceptions and experiences in terms of the meaning they have for a subject. As philosophers Shaun Gallagher and Dan Zahavi explain, "The typical cognitive scientist, on the other hand, takes a third person approach, that is, an approach from the perspective of the scientist as external observer rather than from the perspective of the experiencing subject." There's overlap and integration. Clifford Saron gives nuance to the relation of first and third person designation in scientific research that I cherish: science is done by persons, who describe their crafts in first person subjective terms. The results through unbiased application should result in third person measures, which are then interpreted using first person subjective and second person intersubjective perspectives. And, as he adds, scientists are well aware of the provisional understanding in which the results of their objective research is situated. [10] Certainly, I could feel the Heart in the scientists' dedication. How fascinating, to be part of this endeavor, bringing together our first person subjective experience "from the inside" as participants, in correlation with third person objective science measures, about the same phenomena, information presented in measures "from the outside." I feel privileged to be in this field, where we value the paradigms of both science and spirituality in a way of cooperation, correlation, integration and shared enthusiasm.

An attention test with EEG: first and third person perspective

Let me describe one example of an attention test, performed whilst we were hooked up to the monitor, with 88 equidistant scalp electrodes for EEG and ERP: Electro encephalogram and Event-related potential: the "Visual Continuous Performance Task" (VCPT). This test addresses directed and sustained attention, and has been specially developed for this project. Katherine MacLean and her group report about the outcomes of this test in an article that was published in 2010. Later in this chapter I report about the findings. At this place I just like to address this test more extensively, in first and third person perspective. This "infamous lines-test" we found to be very challenging. While sitting in the corridor next to the research lab, waiting for my turn, some existential musings came up. Here are some diary fragments with impressions of my experiences in waiting and being tested.

Diary, September 11th 2007

> Our scientists are very committed people, handling their subjects with great care and respect. Still, it's an interesting experience to be aware of a very functional aspect in the relationship, in which we participants act, and can be seen as the providers of data; crucial subjects, also to be seen in instrumental context as objects. What kind of respect do real guinea pigs receive? Functioning as a human guinea pig makes me so aware, yes, of the "preciousness of human life." At the same time, the scientists address us as co-researchers, which makes for a nice amalgam of (non) identities. Anyhow, best feels to just be with it, with whatever …
>
> Start of the session: an elastic measuring belt for assessing breathing is applied to my chest. The distance from my eyes to the computer screen is exactly recorded. Brain electrodes are being tested. With electrodes on the chest to check heart rhythm, electrodes in my left hand for skin resistance, a sensor on the

index finger for monitoring blood pressure, I am set. The scientists leave the darkened mobile laboratory and I'm on my own. Cliff Saron's inviting recorded voice sounds through the speakers for control measurements.

Start of the first test. I see alternating longer or shorter vertical lines on the computer screen, shown very briefly, and I have to click with the mouse. First: click on the shorter one. For a start, there are a few short rounds with the feedback of a cheerful tone when I'm right, or a lowering zuffff tone when it's incorrect or when I miss one. I'm told that the distinction between long and short is changing during the task. Indeed, it gets smaller, I feel I need to further sharpen my attention, which makes me aware of subtly stressing muscles around the eyes, and in my shoulders. Relax! Be with it. Later, the instruction changes: now don't click at the short line but at the long line.

After this, so I am told, there will be a somewhat longer round with the short line, without feedback ... Wow, how long is longer? Which turns to be more than half an hour. I feel I lose time sense, and I get distracted more often.

The text below is from the 2010 MacLean et al. article: information about the same events, now as described by the scientists, as third person measures:

> At each assessment, participants completed the threshold procedure (~10 min) followed by the sustained-attention task (32 min) ... In both tasks, participants saw a single vertical line appear at the center of the screen; the line could be either long (frequent nontarget) or short (rare target). The task instructions emphasized the importance of speed and accuracy in responding to the short line ... by pressing the left mouse button.
>
> In the threshold procedure, the length of the short target line varied according to a parameter estimation ... algorithm ..., which determined the short line length that participants could

> correctly detect at a given level of accuracy (e.g., 75 %) *(meaning that each individual participant could detect the target correctly in 75 % of cases).*
>
> Participants received auditory feedback: a ding when they detected a target correctly and a whoosh when they missed a target or responded to a nontarget ...
>
> The ... index of perceptual sensitivity ... was calculated from hit (correct response to a target) and false alarm (incorrect response to a nontarget) rates for ... the sustained-attention task ...

How interesting, the ways of describing the same events! The former presents my personal experiences, first person, and the latter presents data in a way that can scientifically be shared, third person.

In the second round of measurements, like in the first, there was some felt sense paradoxical "struggling to relax" ...

Diary, October 15th 2007

> I hoped indeed that in this second round I would do better and recognize faster whether the line was long or short, and that I would score with less errors, that I would do better and recognize faster when I had made an error, and that I would "recover" quicker. A more relaxed tolerating of my making errors: compassion for myself, and not dwelling in micro-self reproach ... There was this aspect in me of doing, grasping, wanting a good score. For loyalty to the others, the project, for my grasping self that wants to achieve. While over time there was also this surrendering to "non-doing" ...

More attention tests

An example of a well-known attention test was the Stroop Test, showing on the screen words in the colors red, blue, green, and yellow. The request is to stick to domain-specific information: name the color, without being distracted by semantic interference, when, for instance, the word "red" is presented in blue typeface. The hypothesis is that after longer periods of attention training, there will be less slowing down of reaction time due to perceptual-semantic conflict. The experience is that most people don't make many errors; however, they are slower on, for instance, "blue" written in red than when "blue" is written in blue.

Also well-known is the Posner Cueing Task, used in the study of visual attention. We had to detect an image at one of two locations on the screen, while a brief pre-cue was given that indicated the probable location of the image. In general, a correctly cued image is followed by a faster and more accurate response, than when the cue is given at the wrong place. How about the measurements of fastness and correctness after three months of shamatha?

In-between the various tests, sitting at the dark screen, before a new task was presented, we got familiar with the priceless invitation in small white letters: "Rest without engaging in any particular type of directed mental activity."

Blood, spit and tears

For some of you readers: this may be heavy stuff! However, have some patience, as this explanation can't be missed for getting the full picture and the feel of participating in a meditation-science project! Or read through it in a diagonal way. Following are some more examples of third person measurements, while hooked on the monitor, described from a first person perspective.

An example of an emotion test was the Emotion-Potentiated Startle Test: in this classical test we were shown slides that included generally strong affective stimuli: positive, neutral and negative. The negative included, for instance, images of an accident, an amputated body part, a mutilated person, blood and pain. Startle sounds were delivered at specific moments after the showing. The initial hypothesis is that the emotion potentiated startle to negative, compared with neutral pictures, would diminish after training, as well as demonstrate a faster return to baseline state. With these sounds, I experienced myself somewhat pulling together, shrinking, with muscle contractions.

Also, tests were done with film clips: an experimental paradigm of a film viewing task. We saw a film clip, a segment from a documentary depicting graphic scenes of human suffering. For instance, one clip contained shocking scenes from the Iraq war. Afterwards we had to describe the contents and rate the emotions they aroused, with the help of a storyboard of individual frames from the film, arranged sequentially. In this way researchers could create profiles of momentary emotional experiences of the course of the film. Later it became clear that by way of unobtrusively taken video clips, material for the FACS (Facial Action Coding System) had been collected. With these video materials, scorings can be made by science team members proficient in the methodology developed by Paul Ekman and Walter Friesen. The FACS system for classifying human facial expressions is a common standard to systematically categorize the facial expression of emotions. [11] So, in these emotion regulation tests retrospective self reports and facial expressions could be compared. Our first person self-report measures were combined with the third person monitoring of physiological and behavioral variables. This combined approach was innovative: it had not been included before in our kind of meditation research. Hypothesis is that there will be less rejection emotion, anger, contempt, and disgust. Additionally the hypothesis is that there will be less defensiveness in the face of viewing suffering. These were terrible movies, with a lot of physical violence between people, in war scenes. While writing now, I remember the strain and pain of them, certainly in the last measurement, when we were so 'open,' with a combination of sensitivity and resilience at the same time. Some persons expressed their indignation about us being exposed to these clips.

In a different ambiance: hooked up to the monitor, EEG caps on, we have also been evaluated while we were practicing meditation. This was done while we listened to Alan Wallace's instructions in practicing shamatha breathing meditation and in practicing Compassion and Loving-Kindness meditation. It did not feel easy to really relax and meditate with the same mind-set like being quietly on my own, in my room ...

Tons of tubes

To name some more quantitative measures: blood tests included hormone factors (like oxytocin, associated with social bonding and attachment) and proinflammatory cytokine IL-6, giving an indication of immune function (being elevated, for instance, in stress and chronic disease). Brain-derived neurotrophic factor (BDNF) was measured (which is lowered, for instance, in depression).

Certain white blood cells were examined for telomerase. While the telomerase research has been briefly mentioned, here is a further explanation. Telomerase is the enzyme that maintains the ends of chromosomes, the telomeres. It can repair and lengthen telomeres, the sequences of DNA at the chromosomes' tips. These telomeres tend to get shorter at each cell division. When the length of telomeres drops below a certain measure, the cell can no longer divide properly. This leads eventually to a dying of the cell. The length of telomeres is seen as an indicator of cell longevity. It has been shown that levels of telomerase drop down in case of stress.

With telomere and telomerase research, scientists can investigate connections between molecular biology and psychology. Elisabeth H. Blackburn (cell biologist, professor of biochemistry) is a pioneer in this field, being involved in research on stress and cancer. Interestingly: while functioning as one of the scientists of the Shamatha Project, collaborating with Elissa Epel, she is one of the three Nobel Prize winners for medicine, 2009, for her work on telomeres and telomerase! We participants felt so honored with their involvement! [12]

Saliva assessments have been made including stress hormone cortisol, and dehydroepiandrosterone (DHEA). For this, during four days at the measurement periods, four times a day we collected a specimen. Often we could be spotted running to the fridge downstairs, to store them right away.

To continue with this long and incomplete list, the following is about self-reported psychological variables, interviews and questionnaires. We filled in many questionnaires, with psychological and physical health assessments. Among them were a measuring of personality trait dimensions, an ego development scale, state-trait anxiety inventory, depression inventory, surveys regarding mindfulness, self-compassion and well-being, and a questionnaire relating to attachment style.

We were interviewed in half-structured interviews of around one hour at the three measurement periods, recorded on video. Assessed in these were the nature of the goals and expectations that each of us participants had, how these evolved, and in what way they may have been met. Next to that, once every two weeks we made an audio recording about personal experiences.

The Daily Experience Questionnaire that we filled in every evening, enabled the researchers to investigate first person subjective impressions in tandem with data from the third person quantitative measurements. My first person guinea pig experiences included what it was like to be performing lots of attention and psychology tests, letting myself be pierced for huge collections of vials with blood, spitting saliva in tubes, telling about my personal experiences on video, audio, and in diaries.

A memory ...

Diary, October 2007

Yesterday the last measurements of the middle-round. A day with few hours of meditation. Tired in the evening. Catching up. Today a very crisp, sunny day, with 15 cms' snow. Making a walk after putting on high boots. Hardly recognizing the path, hardly any demarcations. Seeing how someone has written in the snow: "Joy," with a heart! I hesitate, shall I jump over it, or walk around it ... I jump-walk around, and then I write, after a few meters:

> "Grace," with a heart as well, readable for a person coming from opposite direction. It arises from the heart indeed. Then jumping, frolicking, skipping. Feels young, old body-memories surfacing. In this context, in this environment: no worries about being looked at, judged, in whatever way.

When I posed questions to the researchers during the project retreat about certain measurements, the aim, and the background, most of the time I received a response like: we can't answer your questions right now, come back at the end of retreat! I remember this last special celebration evening of retreat, when the scientists "at last" gave clarification to us about the set-up of the project, their hypotheses and concerns. Next to the explanations and our shared excitement on this evening and in the last few days of being together, we also shared some wonderings: actually, what happens when you start sitting still for a couple of hours a day ... by itself, this need not influence your philosophy of life, or your insight in the nature of reality. What is the influence of the many variables involved, including in our meditation retreats the amazing natural environment, and the support of being in our group? What really is the intervention? As they noted: so many questions in this field invite for more research ...

1.5 Follow-up and some outcomes

Since retreat, a number of participants from the two groups have continued full-time shamatha for at least several months or even years after retreat. Most of us have continued shamatha meditations at home, albeit for fewer hours a day than during retreat.

Having returned home, after approximately five months, and 16 months after finishing the expedition, we received a laptop containing tests. These were a number of the same tasks that we did during the Project retreat. For the follow-up measurements we were asked to duplicate the circumstances during the Project to the best of our ability: a room as dark as possible, silence, and no disturbance.

We completed the tasks within a brief period of time. Then, after each follow-up round, we shipped the laptops back.

What has come out of this huge data collection, up till now? We participants, of course, have followed the developments with great eagerness. Every bit of information regards "us"! In what follows I'll sometimes speak in the "we" sense, while not exactly knowing, of course, which measurement data have been taken for what outcomes. Yet, the feeling is "we," and of course, in a larger circle, it includes all the scientists, all involved. Outcomes have been published in authoritative peer-reviewed professional journals, and many publications are being prepared that will be presented in the coming months and years. Certainly, the early findings have shown that meditation can have many lasting benefits that range from self-reported well-being right down to the level of our chromosomes. When he reported about the Shamatha Project at the Mind and Life conference with His Holiness The Dalai Lama, in 2009, Clifford Saron elaborated on the initial findings, that demonstrate improvements in adaptive psychological attributes, and in perceptual and attention-related skills. Participants are better in inhibiting habitual responses, meaning that our ability to stop automatic behavior has increased. Alan Wallace writes, in 2009 as well: "Although only a fraction of the terabytes of data gathered in this study has been analyzed ... there is clear evidence that this three-month training resulted in a decrease in afflictive attachment, anxiety, difficulties in emotion regulation, and neuroticism, and an increase in mindfulness, conscientiousness, empathetic concern, dispositional positive emotions, and general well-being."[13]

Transforming the body-mind

Here is an overview of the research outcome articles that have been published in international journals.[14]

Attention and perception

A research article regarding attention and perception, with lead-author Katherine MacLean, was published in *Psychological Science* in 2010. As the researchers note, "The ability to focus one's attention underlies success in many everyday tasks, but voluntary attention cannot be sustained for extended periods of time." They describe the Visual Continuous Performance Task (the "lines test," addressed earlier in this chapter), the 32-minutes sustained attention task, with long and short lines, with pushing the button for the rare target short line. Outcomes of their research on shamatha meditation training have indicated the following: "Training produced improvements in visual discrimination that were linked to increases in perceptual sensitivity and improved vigilance during sustained visual attention." The results suggest that perceptual improvements can facilitate the sustaining of voluntary attention. It was found that we, the meditators, developed greater skill in making fine visual distinctions and sustaining focused attention over a long period of time. After five weeks, we were better able to detect very small differences between the longer and shorter lines. This was also observed at the end of the retreat and in the five-month follow-up in those of us who continued meditating. Additionally, it was observed in the first group of retreat participants, but not our group when the group members were tested as control participants.

Telomerase

Also in 2010, the project team with lead-author Tonya Jacobs published an article in *Psychoneuroendocrinology*. In this article, for the first time, meditation and positive psychological change are linked with telomerase activity. Findings: "Telomerase activity was significantly greater in retreat participants than in controls at the end of the retreat." Telomerase levels were on average thirty percent higher in the group of intensive meditators than in the control group. The meditators who had the highest levels of telomerase coincided with the ones who showed the greatest improvement in some of the psychological tests. Interesting correlations were found. The researchers concluded, on the basis of special statistical modeling techniques, that high telomerase activity was due to the beneficial effects of meditation on perceived control and neuroticism, which by themselves were due to changes in self-reported mindfulness and sense of purpose. Saron, in an interview, remarks in the context of the telomerase findings: "The critical point is that we are demonstrating a relationship between changes in specific psychological traits and telomerase levels that has never been demonstrated before."[15]

Response inhibition

Another study, published in 2011 by lead author Baljinder Sahdra and her group in *Emotion*, showed that we participants became better at a response inhibition. This kind of restraint, in the sense of withholding impulsive reactions to many internal stimuli, including emotionally intense stimuli, seems to be an important factor in healthy emotion regulation. In this research, improvements in the ability at self-regulatory control were found to be linked to better adaptive psychological functioning that the participants reported over the course of the training. An increase in adaptive functioning (AF) means that there has been an increase in mindfulness, empathy, ego resilience and well-being, and a decrease in anxiety, depression, neuroticism and problems in emotion regulation.

As the researchers describe, the capacity for "response inhibition" was assessed in a thirty-two minute laboratory task, with short and long lines with just-noticeable differences.

Instructions emphasized speed and accuracy in making responses to the target, while inhibiting responses to nontarget lines. Recognition, again: yes, indeed, this was what I described earlier, with first person felt sense! So, outcomes of this lines task not only addressed the refining visual perception and sustained attention, but also this response inhibition aspect. It was found that retreat participants, but not the controls, got better at this response inhibition task over the course of the retreat. The controls got better as well, when they later did their own retreat. The improvements were still there at the five-month follow up, showing a lasting change. The researchers note that improvement in the response inhibition task performance during training " ... influenced the change in AF over time, which is consistent with a key claim in the Buddhist literature that enhanced capacity for self-regulation is an important precursor of changes in emotional well-being."

Brain activity

Manish Saggar and his group published an article on EEG findings that came out in 2012 in *Frontiers in Human Neuroscience*. The title is the summary: Intensive training induces longitudinal changes in meditation state-related EEG oscillatory activity. Like most measurements, these were done at beginning, middle and end of the retreat.

Before naming some outcomes, let me connect with my memories that come up when I read this article with the professional language. Also with this language, the article is capable to arouse some felt sense body-mind-soul recollections. Indeed, as is written: at the conclusion of the second day of testing, we engaged in a 12-minute period of silent, eyes-closed mindfulness of breathing practice. We were sitting in the dark, then there was the start sign. The meditation began with around 50 seconds of audio instructions, recorded by Alan Wallace: "During the next 12 minutes, engage in the practice of mindfulness of breathing, focusing your attention on the tactile sensations at the apertures of your nostrils or just above your upper lip. With each inhalation arouse your attention and focus clearly on these tactile sensations. With each out-breath continue to maintain your attention upon the tactile sensations, while relaxing your body and mind, releasing any involuntary thoughts that may arise. So in this way maintain an ongoing flow of mindfulness, arousing with each in-breath, relaxing with each out-breath." And then, to surrender, in this rather strange environment ... certainly, eyes closed was helpful.

Of our 12 practice-minutes while hooked on the machine, 6 minutes of shamatha with focus on the breath were taken for research. For the specialists: the researchers describe how the findings " ... provide evidence for replicable longitudinal changes in brain oscillatory activity during meditation and increase our understanding of the cortical processes engaged during meditation that may support long-term improvements in cognition." The changes were at certain frequency bands: specifically there were replicable reductions in meditative state-related beta-band power, and this was in certain areas in the brain, bilaterally, over anteriocentral and posterior scalp regions: front side central and backside. Also individual alpha frequency decreased across both retreats and in direct relation to the amount of meditative practice (among others, alpha and beta, as well as theta, delta, gamma refer to specific frequencies over brain areas as measured on the EEG). The authors make a link to the MacLean findings about us that show an increase in perceptual discrimination of subtle visual stimuli. On the basis of what they find in the EEG, and other published research, they speculate that intensive meditative practice may also result in increased levels of sensory processing of on-going tactile stimuli. These findings connect with an expanding literature demonstrating functional brain changes associated with various forms of meditation. These changes possibly underlie generalized improvements in cognitive functioning and psychological well-being.

Cortisol and mindfulness

Connecting with the outcomes that she published in 2010, Tonya Jacobs and her research group have presented findings regarding the relationship they found in self-reported mindfulness and resting cortisol output. This article has been published in *Health Psychology* in 2013. Cortisol, produced by the adrenal glands, is often named the "stress hormone." It gets elevated in times of stress and high emotions.

As the researchers note: "Cognitive perseverations that include worry and rumination over past or future events, may prolong cortisol release, which in turn may contribute to predisease pathways and adversely affect physical health." So, this can drive you to illness. On the other hand, "Meditation training may increase self-reported mindfulness, which has been linked to reductions in cognitive perseverations." This is done, specifically, when in meditation we focus on a chosen target, and in that way do not have our attention following uncontrolled, ruminative thought, with cognitive perseverations. As the researchers note, up till now no research has been done that directly connects self-reported mindfulness with resting cortisol. The data generated by the Shamatha Project research (including the fruits of our spitting saliva in tubes for cortisol measurements, and filling in mindfulness questionnaires), provided the scientists the means to investigate this link.

For practical reasons cortisol could not be measured in the wait-list persons during the first retreat. The measurements of all participants in both retreats were combined. In this observational study, the scientists measured self-reported mindfulness (regarding mindful acting, mindful observing, and mindful non-reacting), and p.m. cortisol (afternoon and bedtime measurements) near the beginning and end of our three months' retreat. This is what they conclude: "Mindfulness increased from pre- to post-retreat ... Cortisol did not significantly change. However, mindfulness was inversely related to p.m. cortisol at pre-retreat ... and post-retreat ... , controlling for age and body mass index." (In other words, with mindfulness going up, cortisol was found going down). "Pre to postchange in mindfulness was associated with pre to postchange in p.m. cortisol ... Larger increases in mindfulness were associated with decreases in p.m. cortisol, whereas smaller increases (or slight decrease) in mindfulness were associated with an increase in p.m. cortisol." In the words of the authors: "These data suggest a relation between self-reported mindfulness and resting output of the hypothalamic-pituitary-adrenal system" (this is an important system of hormonal interrelating dynamics). The authors advise replication of these findings in a larger group of meditators, and determining stronger inference about causality, which could be effected when experimental designs are applied with control-group conditions.

An ongoing project

To my and others' feeling the outcomes are quite impressive. And yes, even if the research hadn't told me: it's true for me in felt-sense, that there is more natural mindfulness, more organic purpose in life, less neuroticism (for that, rather ask others around me!). "Perceived control" is referred to in Tonya Jacobs' telomerase article. It is a term in social psychological research … I hesitate, how to relate to this notion, did I feel something in this sense? All depends on how you define it. Comes to mind that Buddhist meditation supports the refining of our seeing where we realistically have some direct control and where not, a practice by itself, in this world of utter interdependence. In this research context the term has been a translation of "environmental mastery," which can be described as: the individual's ability to choose or create environments suitable to his or her psychic condition. This may be seen as a sign of maturity.

Much more to come

The Shamatha Project has become an internationally influential study. The findings and experiences of this project will contribute to fine-tuning and focusing projects to come. More outcomes are to be published, from the data as collected during the project retreat, in follow up after 5 and 16 months, *and: also from a follow up round yet to come, six years after the project retreats.*

The news of a new research grant came in fall 2012. The headline message of the news-release announces that the Shamatha Project has been awarded a grant of 2.3 million dollars of three years to continue and extend this most comprehensive investigation yet conducted into the effects of intensive meditation training on mind and body. The Grant, titled "Quantifiable Constituents of Spiritual Growth" will support the latest phase of research.

In this chapter I have presented information about the set-up of the project's retreats, and about the research measurements, as experienced by a participant. The next chapters are about the meditation practices we learned, back then in 2007, and find their way in some sequential sense up to the present time. While most of this book has this sequential taste, with temporal flow, I have chosen to present the outcomes of the Project research up till now in this same chapter that also described the measurements. It is tempting to disclose more at this place about the new planned round of follow-up! However, I will go with the sequential flow. More about the new follow-up round is to be found at the end of this book, in Chapter 12.2.

This has been an intense expedition, with a long period of preparation, born from a great commitment and sense of exploration in those who brought it into being, for which this guinea pig is deeply thankful.

Chapter 2

Attention meditations

> *The essence of this practice is to attend to the space of the mind and its contents without distraction and without grasping.* (Meditation instructions for Settling the mind in its natural state, 2.3)

After two days of acclimatizing upon arrival, and then two days of running around for the science research, with interviews, attention tests, blood lab, there has been some more space to look into the practices. I've been participating in the collective guided meditation sessions since the first day. After the science fuss, I also got more time to sit in my room, and for settling down into a meditation rhythm.

In this chapter, I continue giving you a taste of the guided meditations that we did, combined with elaborating on some basic notions in shamatha. In the way of presenting I follow the sequence of Alan Wallace's instructions and my own path of getting more deeply involved in the practices.

As briefly mentioned in Chapter 1, and now to be elaborated, we practiced the attention meditations of mindfulness of the tactile sensations of the breath, Settling the mind in its natural state, and Awareness of awareness. Alan explained that the first of these approaches is the primary shamatha technique taught in the Theravada Buddhist tradition, and the second and third are strongly emphasized in the Dzogchen tradition in Tibetan Buddhism. For quite some years before, I had been involved in meditation, mainly in Zen and vipassana. I've practiced Kum Nye Tibetan Yoga, and felt greatly inspired by the books by Tarthang Tulku, among them *Tibetan relaxation: Kum Nye massage and movement, a yoga for healing and energy from the Tibetan tradition*. The books we had at hand during retreat were Alan Wallace's *The Attention Revolution* (2006), and *Genuine Happiness* (2005). This Shamatha Project retreat has been my first more thorough and deep involvement with the Tibetan style of shamatha. Regarding the guided meditations that follow, taken from transcripts of tapes of the sessions: generally, I have chosen one transcript from the first two weeks of the fall expedition. Some elements from later sessions have been added in places for completeness and readability. All the phrasings are Alan's, the terms are his, and the textual flow is his. You may find some repetitions in the meditation instructions; this can't be avoided, and in fact works as a teaching tool in deepening the practice. To me, the repetitions, with slightly different phrasings and intonation have added to nuancing and refining the experience. Try for yourself ...

In concluding this chapter: to complement the description of the science atmosphere in Chapter 1, I add information about some core Buddhist notions, as these nourished our contemplative atmosphere.

2.1 Some basic shamatha notions and instructions

Here, for a start, is a collage of practical descriptions of some shamatha notions as explained by Alan. He often started with inviting us, before venturing into the main practice, to spend a few moments in discursive meditation, meditation including thinking, to bring to mind our most meaningful motivation. Consider this also for yourself. How might you find greater meaning, fulfillment, happiness in your life? Might this be a question of attitude, of attention and presence, or might it include some changes in the way you lead your life? Are there choices to be made? This aspiration may not only regard your individual well-being, but also all those with whom you come in contact, and the world at large. Alan then invited us to find an optimal balance, physically and mentally.

Posture

If the sitting on a cushion with legs crossed, or sitting on a meditation bench was uncomfortable, Alan suggested to either sit on a chair or lie down in the supine position, on your back, with your head resting on a pillow. The last position he connected especially with the whole body breathing shamatha. He often emphasized the importance of relaxation.

If needed, "be your own mentor," and certainly, while meditating for hours each day, find your own balance in shifting sitting in various ways, and lying down. Alan supported what are named the "Seven points of Vairocana," or the sevenfold posture, named after the Buddha Vairocana. This regards Vairocana, the so-called Buddha of the East, meaning "the utterly radiant." The idea is that the posture allows our mind to rest naturally so that the inherent clarity of the mind can come forth. The first point is about placing the legs, possibly crossed, which supports physical stability. It is helpful having the knees lower than the hips. The second is about placing the hands: having them even. This may be with the left hand placed palm up in your lap close to your body, right hand placed in or on the left hand. Or the other way round, if you have grown accustomed to that. Also, placing the hands with palms down on the thighs is fine, right hand on the right thigh and left hand on the left thigh, slightly behind the knees. This is the position Alan used himself. Third, it's important to have the upper body straight, especially the spine: straight, but not rigid. With the body straight, the subtle channels within the body can be straight, and then the energies will flow freely. Fourth, as to the position of the arms: the elbows are not held against the body. This supports the clarity in your mind. Fifth, regarding the neck: slightly bringing the chin back in, and slightly lifting the sternum or breastbone promotes mindfulness and alertness. The tongue lightly touches the palate: this is the sixth point. In this way you don't need to constantly swallow saliva, which may be disturbing. The seventh point regards the eyes, the gaze: in principle eyes can be hooded, just slightly open, certainly with breath meditations. As we saw, with Settling the mind and Awareness of awareness, we keep them open, with a vacuous gaze, not focusing on anything.

Object of mindfulness, introspection

Here I present some basic terminology and summary instructions, as Alan gave them.

Mindfulness, in this context, refers to attending continuously to a familiar object, without forgetfulness or distraction. It is the contrary of mindlessness, the forgetfulness of the object.

The *object of mindfulness* is that object which is selected for the practice of focusing your attention on.

Introspection, meta-cognition, meta-awareness: while the main force of awareness is directed to the meditation object with mindfulness, this needs to be supported with the faculty of introspection-awareness, which allows for the quality control of attention, enabling you to swiftly note when the mind has fallen into either excitation or laxity.

The *object of introspection* is the quality of the attention with which you are observing the selected object of mindfulness. From time to time you will check: is there relaxation, stillness, vigilance? Introspection may show that there is *excitation*: agitation of the body-mind, scattering, often with distraction of the mind. This can be caused by anything destabilizing, like anger, fear, or craving. The antidote to excitation is to relax more deeply. On the other hand, there may be *laxity*: which can be like sluggishness, lethargy, or daydreaming, also referred to as "fading." To counteract laxity, arouse your attention and take a fresh interest in the object of mindfulness.

Dullness refers to a very coarse form of sluggishness, including a lower level of clarity than at the start of the practice.

The instruction, generally, is: to meditate without distraction, or grasping. Practicing without distraction means not allowing your mind to be carried away by thoughts and physical sense impressions, other than the object of mindfulness. Don't let your mind be abducted, hijacked! Practicing without grasping means: no grabbing, no sticking to anything that comes up in the meditation; no preference for one mental object over another, no attempt to control the contents of your mind, and no identification with what comes up.

2.2 Breath meditations, tactile sensations: soothing the body

The focus in these meditations is, respectively: the tactile sensations of the breath in the whole body, in the abdomen, and at the nostrils. The three foci named are more or less diffuse, and the locating of a diffuse focus may feel different from the one that can be rather precisely pinpointed: yet, the focus is clear! And we have the breath with us, all the time.

Regarding the whole body guided meditation, as earlier presented in Chapter 1: I found this a lovely practice, certainly in the beginning of retreat, while I was tired and recovering from jetlag. When meditating in the supine position, with whole body breath meditation, of course we have to balance excitation and laxity, or sleepiness. We cultivate optimally vivid attention, in whatever posture. This supine position is named "the hidden jewel," and I can understand why.

As you read these and subsequent instructions throughout the book, I invite you to allow your awareness to follow their invitations so that you can have a little window into your own awareness. Then rest in 30 seconds of silence before continuing to read, so as to allow yourself a form of embodied reading that supports your own first person experience. It will be helpful to do this following every guided meditation.

Breath in the abdomen – guided meditation

Breathing with focus in the abdomen is familiar to me, from my Zen background. Together with Settling the mind in its natural state: these were and are my co-favorite shamatha practices. Regarding hand position: I've been "brought up" with the special *mudra*, or hand position, with hands supporting each other resting in the lap and thumbs lightly touching one another, the *Dhyana mudra*. During retreat most people, including Alan, had their hands resting on their thighs or knees.

We'll go to the second phase of mindfulness of breathing. We'll focus down here (*pointing to the belly area*). It can be very helpful in stabilizing, calming and grounding the attention. Let your awareness, your attention come and settle down in the sensations of the abdomen, with the expansion and contraction of the abdomen with each in and out-breath. Direct your attention there with your best approximation of bare attention, quietly, silently witnessing the sensations with as little conceptual projection, superimposition as possible. Gently arouse your attention with each inhalation, thereby overcoming laxity; and with each exhalation release any involuntary thought, relax your mind and body. Maintain an on-going flow of mindfulness engaged continually with this on-going flow of the tactile sensations with the rise and fall of the abdomen.

You may find it helpful to experiment with counting of the breaths. What I would suggest is one very brief count, staccato, a mental count at the very end of inhalation. Release all the way to the out-breath, releasing any involuntary thoughts. Then quietly arouse your attention through inhalation ... then the second staccato count, "two" ... and as we count 1 through 10, what you're doing is thinning out discursive thoughts, replacing many involuntary thoughts with a few voluntary ones, milestones on the path of the in and out-breath.

The focus of mindfulness, the object of mindfulness is the sensations in the abdominal region correlated with the in and out-breath. With the faculty of introspection, monitor the meditative process as always. As soon as you see that laxity has set in, that you're getting a bit dull, that the clarity, the vividness is tapering off, then freshly arouse your attention especially with inhalation, balance out. And in an even more likely event, when you find yourself caught up in thoughts, carried away in this practice, immediately relax, release and drop …

Let's bring the session to a close.

As Alan stated, in relation to counting: it's good to bear in mind that the practice is mindfulness of breathing, not mindfulness of counting! When counting was helpful, his advice was to practice it just intermittently to firm up one's mindfulness, or count not at all if it simply cluttered one's mind. To me, it has felt helpful to include the counting for a few minutes at the beginning of a session. Sometimes, when my mind was very busy, with quick distraction from focus, I did a few minutes more later in the session; with a sense of more decidedly binding the attention to the object.

Regarding freshly arousing your attention: Wallace taught us to heighten our interest in the object of mindfulness in case of laxity, and in general for refining attention, with applying our interest to greater detail. This refers to detail in the beginning, the trajectory and ending of the breath movements, specifically on in-breath. With unrest and agitation we were invited, time and again, to relax and emphasize longer out-breaths.

Breath at the apertures of the nostrils – guided meditation

After whole body and abdomen as a meditation focus, coming from more or less diffuse focus, this next shamatha breath approach has a highly specific focus. Feel invited to join.

Today we'll touch the third mode of mindfulness of breathing, focusing up here, at the nostrils. Let this initial phase be settling deeper and deeper into a sense of physical and mental ease. And to the best of your ability, as you then move into establishing stability and vividness of attention, the sense of ease and looseness of the body-mind do not diminish.

And now let's elevate and narrow the focus of attention to the sensations of the in and out-breath to the apertures of the nostrils or above the upper lip, wherever you most distinctly perceive the sensations of the breath. Place your mindfulness at that aperture as if your awareness were a gatekeeper at the gates of a walled city; not following the travelers inside a city or outside, but monitoring just the gateway. Your mental awareness in this case conjoins with your tactile awareness. See that your visual attention, that your eyes themselves are disengaged from the practice. See to the best of your ability if you can maintain an on-going flow of mindfulness, as there is an on-going flow of sensations, both during in and out-breath, but also during turn-around points. Savor the sweet spot at the very end of the out-breath, and then effortlessly allow the in-breath to flow in. Let this be a full-time engagement, with no time off, no lapses. When counting, do not let this turn into mindfulness of counting, which is quite crude. It will not take you very far in the path of shamatha ...

With one corner of the mind, one facet of your awareness, arouse your faculty of introspection to monitor the quality of mindfulness itself, to see whether it is lost and you're simply caught up in thoughts and forgetfulness, mindlessness, which is the opposite of mindfulness. In which case, immediately release whatever thought has carried you away, and return immediately to the sensations of the breath. But do so with relaxation.

Let's bring the session to a close.

... And allow yourself 30 seconds for integration, before reading on.

A remarkable aspect of this breath meditation on the apertures of the nostrils is that the further you progress, the subtler the breath becomes. It may at times be so subtle that you can't detect it at all; indeed I sometimes wondered about that. This, according to Alan, challenges you to more and more refine the vividness of attention. You have to pay closer and closer attention to these sensations in order to stay mentally engaged with the breath. This refined practice of mindfulness of breathing has a unique natural biofeedback aspect, in the sense that you gain greater awareness of a physiological function, with the instrument that provides information about the activity of that function. There is a biofeedback loop: hardly feeling the breathing stimulates the sharpening of your awareness for that feeling, with a growing refinement in it.

Regarding "with one corner of the mind," one facet of your awareness, arouse your faculty of introspection to monitor the quality of mindfulness: this monitoring aspect, as Alan stated, is a "little corner of insight practice," as part of the calm abiding practice.

2.3 Mind: the space of the mind, and what arises in it

In the beginning of the retreat I had the feeling that there was a lot to practice for me, with the three breath approaches outlined here. I could easily be with them for weeks, or months. However, the cineac-carousel just turned its rounds, and after one or two days of each breathing practice, we continued with the guided meditation on Settling the mind. Settling the mind in its natural state may be seen as a general expression, referring to an important outcome in long term robust shamatha practice. Next to that, the expression stands for this specific shamatha practice that we did during the project retreat: this practice focusing on mind and mental events. In the present book most of the time Settling the mind in its natural state will refer to the specific meditation practice. In the subsequent chapters I'll explore this practice from various perspectives. In Chapter 6.1 the instructions by Lerab Lingpa, a nineteenth-century Dzogchen master, are presented. Alan Wallace told us that they form the basis for his instructions.

Settling the mind in its natural state – guided meditation

The following shows how Wallace, after some introduction by way of the breath, guided us in the meditation on the mind. These have been the first instructions I heard, for doing this practice.

Let us settle the mind in its natural state. In this practice the object of mindfulness is *the space of the mind and whatever arises in it.* In this practice, we let the eyes be at least partially open and the gaze resting vacantly in the space in front of you. We do not focus on any visual object, any shape or color, just rest the gaze vacantly. And now, direct the full force of your mindfulness to the space of mental events, events or contents of experience that cannot be detected with any of the five physical senses or any of the instruments of technology, the space in which discursive thoughts arise, mental chitchat, mental images, memories, fantasies, desires and emotions, all manner of mental events. Direct your attention to this domain of experience, not included in any of the five physical senses.

Now quietly observe whatever next arises within this field, this space of the mind, be it another discursive thought, perhaps a mental image. Whatever arises, just let it be as you attend to it with discerning mindfulness, in no way trying to reorder or modify the contents of the mind. The space of the mind doesn't have any particular location in front or behind, down, above or below. It is simply a domain of experience, the one that is left over beyond the five physical domains of experience.

The essence of this practice is to attend to the space of the mind and its contents *without distraction and without grasping.* Distraction occurs when we are carried away by the thought, into the past, imaginary future, or into thoughts about the present; or to the referent of the thought. Grasping occurs whenever we latch onto mental events as "I" or "mine," identifying with that which arises; or respond to them with preference, liking this, disliking that. Attend to it for what it is: an event arising in the space of the mind, with no owner, no controller. No preference also includes not having a preference for the thoughts to dissipate and dissolve back into the space of awareness. Be as indifferent as space itself, nonreactively attending toward whatever arises, illuminating its nature but without intervention. Whatever arises, be it pleasant or unpleasant, wholesome or unwholesome, long or short, subtle or coarse: let it be, and observe the nature of whatever arises in the mind with discerning mindfulness.

Bring the full force of your mindfulness, the focus, the attention, to the domain of experience that is purely mental in nature, selecting out all the appearances that arise to the physical senses. Allowing them to arise, of course, we have no other option; but not paying attention, not deliberately focusing on, not taking an interest in any of the five physical sense fields. In many cases you may be aware of the thought, the image, certainly the emotion, only after it's arisen; it may have gone on for some seconds before you're aware of it. But as soon as you become aware of it, simply direct your mindfulness there, as you release the grasping. Just let the thought, image, whatever it may be, just let it arise with as little entanglement as possible.

(Later, the instructions included: shift the attention from that what arises to the space of the mind). Rest in the larger space of your awareness as you attend to the subspace of the mind. As soon as you know that your mind has been caught up in excitation and distraction, agitation: release not the thought but your grasping onto it, and relax more deeply while letting the thought be, without banishing it. Unlike other shamatha practices, here we do not banish or let go of the thought, we do not even prefer for there to be fewer thoughts rather than more. Rather simply release the grasping onto the thought and observe whatever thoughts arise in the mind with unwavering mindfulness. Exercise stability and wakefulness as qualities of the subjective attention, not of the objective content of the mind. That comes and goes, and is sometimes very chaotic. You exercise the instrument of focused attention …

Let's bring the session to a close.

Why the instruction of leaving the eyes at least partially open, in this practice? With this, Alan said, the artificial barrier between inner and outer begins to dissolve. In the beginning I found this was not easy, certainly not when there are many objects (persons, things, movement) in view. However, in my own room I found a good position. Sitting before a blank wall has been quite helpful for a while. Also, lying on the bed and viewing the ceiling sometimes felt like the right thing to do.

Alan instructed the meditator to direct the attention to the domain of experience, not included in any of the five physical senses. In Buddhist psychology, next to the five physical senses (vision, hearing, tactile sensations, tasting, smelling), the workings of the mind are seen as the sixth sense, the mental sense.

For those who have some problem with finding the space of the mind, in the beginning, Alan advised to deliberately generate a discursive thought. This could be any thought whatsoever, for example, the thought "this is the mind." Slowly and deliberately generate this mental sentence, syllable by syllable, and while doing so, direct full attention to this mental event. As soon as that thought has vanished, back into the space of the mind, keep your attention fixed right where it was; then you have found the domain.

We have often been invited to shift the attention, from focusing more on the foreground, on that what arises, to more on the background, the space of the mind, and to rest in the larger space of our awareness while attending to the subspace of the mind.

This metaphor by Sogyal Rinpoche, contemporary Tibetan Buddhist teacher, has been regularly quoted: "Your awareness is like a gracious host in the midst of unruly guests." [1] Guests come and go, they may quarrel and fight, but the host remains calm. Certainly in the beginning, my subjective awareness was mainly with the objective phenomena arising in the space of the mind, and sometimes completely identified with thought trains and persisting memories. With practice, mindfulness gradually becomes more stable, and one grows more familiar with the space and the arisings, even when there are many "unruly guests," when there is mental turbulence. In fact, nothing can harm the mind, whether or not thoughts have ceased, whatever mental content comes up. As Alan said: a Tibetan word for meditation is *gom*, which includes the meaning "familiarizing." Yes, it feels like getting more familiar with these arisings, and the processes in the space of mind. Even when there is much agitation, as one is not distracted and not grasping, awareness remains still. This is called the "union of stillness and motion." A guided practice with this name will be presented in Chapter 9.2.

Diary after retreat, 2009

> Sitting behind the laptop, putting in an audio disc of a meditation of two years ago, organizing myself in some posture of mixed relaxation-with-one-hand-with-pen, writing ... While listening again to these instructions on tape, then writing down, I'm so aware that a few years ago during retreat, these same words carried quite a different meaning for me in comparison to now; and also that they landed in quite a different mindscape than there is now. Is there a nostalgia for the freshness of the experience, present at that time? The feeling now: there's more relief, more context and depth. What is it that has changed? The meditator, the practice, the way of practicing Settling the mind in its natural state? The mind, the brain, the context? The answer is: all of them. Still: freshness, again and again!

2.4 Awareness: shamatha without a sign, illuminating awareness

A few days after starting with Settling the mind, instructions proceeded to Awareness of awareness.

Awareness of awareness – guided meditation

Here is an example of a guided meditation in Awareness of awareness, also named "shamatha without a sign," or "quiescence without a sign," without a specific focus. So, in that sense this meditation is different from the ones presented before, as these all had a focus:

Let's begin by settling the body, speech and mind in their natural states and the breathing in its natural rhythm.

Let your eyes be at least partially open and let your gaze rest vacantly in the space in front of you. Now with each in-breath, *invert your awareness right in upon itself*. Draw your attention right in upon that which is observing, vividly, rigorously. Attend to the very event of awareness itself, taking nothing else as your object. With each in-breath, draw your awareness inward upon that which is observing; with the in-breath, arouse your attention, pay close attention, in this way overcoming the attentional imbalance of laxity. And with each out-breath, utterly let go, release your awareness out in the space in front of you, taking nothing as an object. Just release, with no object, out into space. Just rest. Let awareness rest in its own state, aware of the on-going flow of being aware, taking an interest in nothing else. Nominally, the "object of mindfulness" is awareness itself.

Invert your awareness in upon itself, without grasping onto any subject, which is nowhere to be found. Whenever your attention latches onto an object, whenever an involuntary thought or image arises, immediately *release both the thought and the grasping* and rest your awareness in its own nature.

It is said that the salient qualities of consciousness are luminosity and cognizance – the sheer event of knowing. Are these qualities of your own awareness?

Shamatha without a sign is in some respects the subtlest of all the shamatha practices we're emphasizing in this retreat. In fact, this can be the most relaxing of all practices. You focus your attention not outside or inside, not on an object; awareness is as well inside as outside, and precedes this differentiation.

Let's bring the session to a close.

Again, rest in silence for 30 seconds before continuing.

Nominally, in this practice the object of mindfulness is awareness itself. You could say that awareness takes itself as its object. But experientially, as Wallace explained, this practice is more a matter of taking no object. You simply let your awareness rest.

The instruction is: "Invert your awareness in upon itself, without grasping onto any subject, which is nowhere to be found." There's no subject, no "me" around. Who's meditating? We'll come back to that.

As is mentioned, " ... whenever an involuntary thought or image arises, immediately release both the thought and the grasping." So, this instruction differs from the one in Settling the mind in its natural state, where just the grasping is released, and where the thought is not banished but observed.

Alan suggested that in the beginning, we may approach this practice in a preliminary mode that can make it more accessible, by very gently conjoining it with the breath. A very gentle pairing of shamatha without a sign with the breath, and engaging in this undulating movement is an approach that Padmasambhava taught. Certainly, it is a practice with quite an honorable and lengthy past, as Padmasambhava, the great Indian meditation master who came to Tibet, lived in the eighth century. This practice has been taught for centuries since then, as a way of overcoming both laxity and excitation in one breath cycle. In the beginning, I found this conjoining with the breath very helpful. Later, to me it felt a bit like staying anchored to the body, in a way that might preclude the "just awareness."

Wallace noted that in following this method, we do not let the mind go blank! We have to ascertain the absence of ideation as the meditative object. The concern is that the meditator should not float into some trance, but maintain an actively engaged focus and introspection throughout this practice.

2.5 Some shamatha questions and answers

At this point, let me give an impression of what came up in the afternoon exchanges: some questions and answers, and some contemplations and notions that played a background role, day-today, during the project. Of course, it is a personal selection. In these sessions, the only hour of talking during the day, we got to know each other in our experiences, wonders and worries …

Ascertaining awareness, stability and vividness of attention

As we were getting on our way, quite a few exchanges occurred about how we experienced the refining of perception and attention. Unavoidably, for a start we became aware of the astounding busyness, turbulence and chaos, this feeling of over-crowdedness in our minds. Also, of the way that we most of the time came behind the facts, only seeing how we had been distracted, when some time had elapsed. The first skills one has to learn are: registering that one has been distracted, and then redirecting attention to the focus.

Alan elaborated on the Buddhist psychological view of six types of consciousness, connected to the six sense modalities: the five physical senses – seeing, hearing, touch, smell, taste – and the sixth sense, of mental perception. When we detect something by any of our six senses, there is a short moment before the mind projects concepts, labels and classifications onto our immediate experience. To discern this brief moment demands a high degree of vividness.

Why is it important to make this discernment?

The importance of this brief instant is that it is an opportunity for gaining a clearer perception of the nature of phenomena, including a subtle continuum of mental consciousness, Alan explained. It is this continuum from which all forms of sensory perception and conceptualization arise. In order to be able to get to a clearer perception of the nature of phenomena, it's crucial to see where conceptualization, personal bias and elaboration come in.

According to Alan, it is stated by an important school in Buddhist psychology that there are about 600 pulse-like moments of cognition per second. This would refer to some 2 milliseconds per pulse. Roughly, this is in agreement with modern psychological insights.

We can imagine frames in a motion picture, giving the impression of a continuum. In that sense we rather speak of a process than of mind-moments with a duration.[2] However, generally we are conscious of far fewer moments than these named 600 per second. Buddhists speak of the moments of cognition that we are not aware of as "non-ascertaining awareness." Appearances arise in the mind, but we don't register them. Afterwards, there's no notion of having witnessed them. When we are in a shop with people, products, and background music, it's clear that many sensations are presented to our awareness, but only a fraction of them is noted. Our attention is very selective.

How do these mathematics relate with our practice?

In connecting mathematics with practice aspects of relaxation (1), stability (2) and vividness of attention (3), the following clarifications have been very helpful to me:

Practicing shamatha includes a lively sense of relaxation (1). Relaxation is a precondition for cultivating more ascertaining awareness: with more and more moments of these six hundred pulses per second, that, potentially, can be ascertained.

Attentional stability (2) is a measure of how many of the ascertained pulses of awareness are focused on the object that we have chosen. This is about *coherence* and continuity of the mind moments with awareness, with regard to the chosen object. So, when, say, 50 moments out of 600 are ascertained awareness, and 45 are on the chosen object (the breath, for instance), this is a relatively high rate, with good continuity, meaning that there is little distraction. This means: only five of the ascertained awareness moments are not on the focus, showing that staying on the focus is relatively complete. In a case, for instance, when 20 out of 50 moments would not be on focus, there would be more of what is named "partial staying." As we can experience: attention may withdraw from the meditation object, to various degrees. Sometimes involuntary thoughts take the center of our attention, while the meditation object is displaced to the periphery. After more practice, we can have the meditation object at the center, while some noise may remain at the periphery. In that case, the staying is less partial, and more complete. Complete staying would be when the attention is at the meditation object at all moments of ascertained awareness.

Attentional vividness (3) is a measure of the *density* of the ascertained moments of awareness per second. With more moments of ascertained awareness, the density is said to increase. For instance, this might go up from 50 (as in the former example), to 100, per 600 pulses. With more advanced practice, increasingly more mind moments of ascertaining awareness will be on the chosen object, with increasingly more continuity and more completeness.

I felt struck by Alan's elaboration on how many meditators (and here I also recognize myself, sometimes) especially like to increase vividness. This can bring some feeling of intensity, of "high." But, for vividness to be lasting, relaxation (first) and stability are the great prerequisites. Otherwise your vividness will be like "writing on water." It was emphasized also that, underlying these we need a foundation of equanimity, as a general sign of spiritual progress: imperturbability, in presence and attention, while confronted with the turbulences of life.

Development, discovery and grasping

What I found an interesting theme, linking with describing the process in terms of stages, regards developmental and discovery aspects in the path. Alan Wallace explained that many modes of shamatha are designed along a developmental model of cultivating stability and vividness. There is also the discovery aspect that evolves specifically in increasingly refining stages of mind and awareness practices: the unveiling or discovering of the innate stillness and luminosity in the very nature of awareness itself. An unveiling by releasing all grasping to subjects and objects: this is discovery, or realization, of that which has been there all along.

I wondered about the way of relating between these aspects of development and discovery, especially also in Settling the mind practice. What is their relationship with grasping? While the developmental aspect counts for all the shamatha meditations, after prolonged practice this aspect feels like less and less on the foreground for Settling the mind and Awareness of awareness. It feels like there are states of awareness that are just there, where indeed, with "dust" decreasing, we may be in contact with that which has been there all along. The breathing practices have a more clearly developmental linear feel. Yet, for all of the shamatha meditations it's there. We practice and practice, and gain more refined abilities and perceptions in mindfulness and monitoring. The development aspect in the practices does contain an aspect of grasping, even if, after long term practice, this may be very subtle grasping.

How subtle can this grasping become?

As Alan explained during one of the afternoons, with a focus on the breath, some people find there's a lot going on, creating some tightness and tension. Even in the practice of Settling the mind in its natural state there's still an object you're grasping onto. It's very subtle grasping when the practice is done correctly, but nevertheless, you're reaching out and touching something, and then holding on in a manner of speaking, very lightly. In the practice of shamatha without a sign, it's really only a marginal level of grasping, because when you draw your awareness inward, there's nothing there to hold on to. Indeed, I've very much experienced fascinating differentiations and subtle degrees of these body-mind tensions and tightness. Discovery comes in, according to Alan, when one gets more insight in the working of the relative mind, including its grasping tendencies.

As Alan noted, Padmasambhava has commented that the practice of shamatha without a sign may in some rare cases be sufficient for ascertaining the nature of pristine awareness. This is a practice of profound inactivity. One is *being* aware of being aware, but one is not really *doing* anything. The illusory separate-self sense temporarily has nothing to do. For exceptional individuals with "little dust on their eyes" this may be enough for fathoming the ultimate nature of the mind and its relation to reality. [3] Indeed, exceptional individuals, with so little dust on their eyes ... However, as was clearly underscored by Alan, while relative insights, connected with the subtle continuum of mental consciousness can be ascertained by means of the cultivation of shamatha, pristine awareness and ultimate insight is usually realized only through the cultivation of vipashyana.

Substrate consciousness

Is there any endpoint in shamatha training?

The cultivation of shamatha is widely known as a means of ascertaining the named subtle continuum of mental consciousness, also known as the "substrate consciousness," the relative nature of the mind. According to the Tibetan Buddhist view, ascertaining the relative nature is the most natural or settled that our human mind can become. With shamatha practice we can attain substrate consciousness, with a mind characterized by great suppleness, pliancy and dexterity, a mind optimally healthy. So, Alan's response here is: yes, there is an endpoint in the shamatha training. Substrate consciousness is the ground of the (habitually confused) mind or psyche. The psyche can be seen as the totality of the human mind, conscious and unconscious. In attaining substrate consciousness, the mind without content has been reduced to its bare nature. In this state the mind is bound by time and causality, and specific to an individual. When, in this state, the mind is withdrawn from the external world, it is qualified by distinct experiences of bliss, luminosity, and nonconceptuality. Yet, shamatha is not a once for always achievement; one has to keep up practice, otherwise focused attention will deteriorate.

In Settling the mind in its natural state shamatha practice, while addressing the relative ground of the psyche, many insights about the workings of the mind arise.

Diary, September 2007

> *So* many thoughts, trains of images, plans, memories jumping over each other, cascading. I'm *so* tired of this agitated way of "doing my best," I'm bumping into this tendency also here, getting to see patterns as if by microscope ... old traits, with defensive compulsiveness – as if some part in me needing to protect itself, in unfamiliar circumstances? Feeling a subtle straining in my body, in my hands-muscles, neck, intestines.

Insights in the mind and in mind-body processes bring up the question: what *is* mind? This is a really good question for vipashyana practice! Alan found out that it is very helpful, when teaching shamatha mind practice, to include one day of vipashyana with questions about the mind. What is the mind, and where is it located? Does it have a shape, a color? Does it have boundaries? This little bit of vipashyana about "what the mind is" skillfully complements the "insights" gained during shamatha mind practice.

Nine stages of meditative concentration

With increasing relaxation, stability and vividness, contours of a path with stages show up. This path has been walked by numerous contemplatives through the centuries, and has been mapped in a trajectory of nine stages. A few practical words about these, as they were presented during the expedition, may be helpful at this point. An illustrative representation of these stages is often referred to as "The Elephant Path." The image, and more theoretical context will follow. What is the meaning of these stages? How is recognition of the stages helpful in the practice? The nine stages, as described by various authors through the ages, are meant as markers on the path. They are not meant for measuring or even judging oneself and others on them. At this point, next to naming the nine stages, let me give an idea of three practical aspects in each stage that have been valuable for many, including me. They give some orientation of "where one may find oneself," while on expedition. The following overview is based on what Alan Wallace told us about these stages. [4] I add a few comments of my own. While we won't go into explaining all the details, with even the briefest attempts to practice, you may recognize some of these aspects right away. The information, presented per stage, regards specifically: 1 "What is achieved," 2 "The power by which it is achieved," and 3 "What about involuntary thoughts?"

1. Directed attention

1 You are able to direct your attention to the chosen object ... even for a second or two (this is really positively formulated!)
2 Learning the instructions
3 The flow of involuntary thought is very busy ... often compared to a cascading waterfall (in reality: beautiful to look at, while in meditation metaphor: confusing, exhausting ...)

2. Continuous attention

1 There is attentional continuity on the chosen object up to a minute
2 Thinking about the practice
3 The flow of involuntary thought: like a cascading waterfall

3. Resurgent attention

1 Quick recovery of distracted attention, attention is mostly on the object (for sessions of 30-60 minutes)
2 Mindfulness, non-forgetting the object of attention
3 The flow of involuntary thought is (still!) like a cascading waterfall

4. Close attention

1 You no longer completely forget the chosen object (which may

give a feeling of relief, a sense of accomplishment, maybe some complacency)

2 Mindfulness is now strong

3 The stream of involuntary thoughts goes a bit more quietly: they are like a river quickly flowing through a gorge

5. Tamed attention

1 You take satisfaction in samadhi (the challenge is: while now being free of coarse excitation, not to succumb to laxity)

2 Introspection

3 Involuntary thoughts are like a river quickly flowing through a gorge

6. Pacified attention

1 There is no resistance to training the attention (the challenge: coping with emotional upheavals, like fear, and feelings of depression)

2 Introspection

3 Involuntary thoughts are now like a river slowly flowing through a valley

7. Fully pacified attention

1 There is more ease, with pacification of attachment, melancholy, and lethargy

2 Enthusiasm

3 Involuntary thoughts are like a river slowly flowing through a valley

8. Single-pointed attention

1 Samadhi is long, it can be sustained without any excitation or laxity (attention is very highly focused)

2 Enthusiasm

3 Increasing quiet: the conceptually discursive mind is calm like an ocean with no waves

9. Attentional balance

1 Flawless *samadhi* is long, sustained effortlessly. It can last for at least four hours
2 Familiarity
3 The conceptually discursive mind is still.

Connecting with these stages of refinement, there are increasingly subtle forms of excitation and laxity. To give you an idea: Wallace explains "coarse excitation" as the attention completely disengaging from the meditative object. In "medium excitation," involuntary thoughts occupy the center of attention, while the meditative object is displaced to the periphery. With "subtle excitation," the meditative object remains at the center of attention, but involuntary thoughts emerge at the periphery of attention. In the movement there is gradually less partiality, and more completeness of attention on the object. Regarding laxity: in "coarse laxity," attention is described as mostly disengaging from the object due to insufficient vividness. In "medium laxity," the object appears, but not with much vividness; and in "subtle laxity," the object appears vividly, but the attention is slightly slack. There's a lot of work to be done. Alan emphasized: doing shamatha is a full-time job. As I understand this: practicing in a proper way requires all our attention, at any moment, on the way to more continuity and completeness. To hear about this trajectory with many details, and about the many months, or even years of diligent practice that it might take, made quite an impression on me. It points to the fact that in the general Western meditation scene three months sounds a lot, while in this context of fully achieving shamatha, three months is "peanuts" ...

Is it possible to skip one stage?

As Wallace explained: these stages are sequential. You can't jump over one to the next. The path starts with a mind that cannot focus for more than a few seconds and culminates in a state of sublime stability and vividness that can be sustained for hours. The meditator progresses through each stage by rooting out progressively more subtle forms of the two obstacles: mental agitation, and dullness or laxity.

For didactic reasons, and for fitting in a daily structure, our shared sessions in the shrine room were 24 minutes, in general, one *ghatika*. This was considered in ancient Indian and Chinese theory to be the best duration for practice when you start meditating. It is said to be the duration for vital energies, the *qi* in Chinese, in the body to go through one complete cycle. In our practice in our own rooms, we gradually made sessions longer.

Which practice to do?

This question often came up: while having addressed various flavors of shamatha now, which practice to chose? Alan gave some general recommendations: for the first four stages, he suggested to practice whatever method we found easiest. Still, for achieving the first four stages, he advised the practice of mindfulness of breathing. By stage five, the mind is relatively stable, and one can move on to subtler techniques. By focusing on an object of any of the physical senses, you can certainly develop good relaxation and stability. However, for vividness you won't get to the full potential. Sensory awareness in a way is too coarse for this. A mental object is needed. When reaching more advanced phases of the shamatha practice with mindfulness of breathing, practicing the most refined form of breathing meditation directed at the apertures of the nostrils, a sign will come up, a so-called "acquired sign." This is seen as a symbol of the air element that appears for the mind's eye. To different people this sign may appear like a star, a wheel, a cluster of flowers; some round pattern. As soon as this mental sign arises, one needs to leave the physical focus and shift the attention to this mental sign. [5]

Beginning with the fifth stage, Alan recommended to either practice Settling the mind in its natural state or Awareness of awareness. Certainly from the eighth attentional stage and onward, we move on to the still subtler practice of maintaining awareness of awareness itself. The practice here, as has been elaborated, is not so much one of developing attentional stability and vividness as it is of discovering the stillness and luminosity inherent in awareness itself. Clearly, on this expedition we didn't stick to these basic recommendations. As Alan wanted to teach us all the named shamatha forms, right from the beginning, independently of what stage we were at, we got to practice all of them.

In the weekly private interview each of us discussed with Alan what practice to emphasize, and how to continue. For the first few weeks, also for me, it was certainly helpful to include a good amount of breath meditation, to ground and come home in some way. Still, the choice of which meditation to take as our main practice was left up to us individually. It has been Alan's experience, he explained, that many people find the practice of Settling the mind in its natural state to be rather challenging, while others take to it naturally. The practice of Awareness of awareness is subtler still, and is taken up by many in later phases, but in his view it may be optimal from the beginning for those who are strongly drawn to it. The healing aspect in the practices, in relation to the psyche, is there anyhow. With Settling the mind healing happens as if "under a transparent bandage": you observe the turbulent emotions and you may see them "melting" in some way. With breath and awareness practices it's rather as if happening under a closed bandage. With these, emotional turbulence may rather show up in between sessions than during the sessions. Anyhow, any one of the three main approaches (body, mind, awareness) does bring progress along all stages of attentional development.

I remember one friend deciding to mainly do the breathing practices for the three month duration of the retreat. He was quite happy with it and could calibrate his progress well. Regarding Settling the mind practice, some participants shared a dislike of this meditation. They experienced it as complex, or "bringing up stuff," or "keeping them so much connected with thought," which they didn't like. There were also others, like me, who enjoyed Settling the mind. Some just fell for Awareness of awareness. For me, breath meditations have taken the lead in the first month, while Settling the mind in its natural state became my second main practice later. Still, during retreat, every day I practiced forms of all three: including body, mind and awareness focus.

Nyam

Alan Wallace has explained to us the phenomenon known in Tibetan Buddhism as *nyam*. It refers to a transient anomalous psychological or somatic experience that is catalyzed by the practice of meditation. He gave this information in the beginning of the expedition; probably finding this beneficial, warning us in advance ... *Nyam,* often translated as "meditative experience," or meditative realization, can be experienced by the meditator as pleasurable or unpleasurable, as good or bad. As Alan kindly announced: *nyam* (the term refers to both singular and plural) are bound to come up. It is not that they are created by the meditations; they are uncovered. *Nyam* comes up most explicitly in Settling the mind practice, where mind-body correlates, and energetic dynamics are strongly felt. One may experience intense anxiety, sparkling joy, beautiful visions, deep cravings, and all kinds of ups and downs. The basic instruction, of course, remains: just be aware and observe, without distraction or grasping.

Nyam may give the meditator a rough time. How to handle these strong emotions, anxieties, and confusions? Some of Alan's advice has been:

1. When needed, consider using the supine position: this general "infirmary" can be very helpful.

2. Walk outside, to places with great, broad views and perspective. He named Marpa point as such a place, a nearby mountaintop (named after Marpa, an important Tibetan Mahamudra sage living around the eleventh century CE).

3. Try broadening, widening the meditation. For instance, in Settling the mind, when the content of the mind is turbulent, one

may focus on the background, where awareness is quiet, serene. Do not ruminate, with narrowing the mind. This is tunnel vision. Tunnel vision is really different from shamatha: while shamatha is highly selective, it is not narrow. Shamatha is selective, even with a very broad object of mindfulness, like the space of the mind with contents, or just awareness! So: widen the scope ...

4. Do one of these two: either, cut right through, going forward, advance: go right away to practicing Awareness of awareness. There is just stillness, you have "no interest" in what else is there. Or, alternatively, retreat. Go to breathing practice, with a clear focus, if that is of more help.

5. Another practical possibility has been to drop a note on the message board for Alan, who might then address the subject in the evening meeting. (An example of instructions for widening the meditation scope is given in Chapter 7, an example for advancing and retreating is given in Chapter 4).

Wallace emphasized that, with physical phenomena that may represent medical problems, we don't take risks. There is a nurse around, and a doctor can be consulted, which has indeed been done a couple of times during retreat. Otherwise, when there is *nyam*, then just watch your mind heal. So, his care was clear, and we were all set. It certainly made me curious about what might come up.

While in Chapter 5, with extensive diary notes, more will be revealed about turbulences and turmoil, for concluding this chapter I want to present some basic notions from Buddhism. These will support you in staying attuned to what further unfolds.

2.6 Some basic Buddhist background

In what has been presented in Chapter 1, the sphere, taste, language, agency and culture of Western science, with all its impressive accomplishments, shine through. Still, clearly, additionally, everyday during our participation in the Project there was this bathing in the field of disciplined practice, with Alan's guided meditations, enriched with his explanations, responses to our questions, and his many stories from his personal experience. Certainly, the practices could be done, in some sense, independently of place, time, and culture. This made them well suited for research. On the other hand, the practices had a history, a cultural context, and the setup of this project did not exclude these. Some (a little) skillful reading, directly relating to the practices was stimulated. As is evident from the questions and answers section, we were taught in a way that brought in the Buddhist touch.

Buddhism, one may say, strongly connects with first person approach, and we, project participants in the field of meditation and science, were stimulated by Alan to explore, and in the afternoon to share our direct experiences. We received the basic Buddhist invitation, to "really see and experience for yourself," as is presented in the *Kalama Sutta*. The Buddha's propositions relating to meditation, psychology and the nature of reality – see below – can be taken as hypotheses to be verified or falsified. In his teachings to the Kalamas, about investigating and experiencing for themselves, the Buddha encourages them: " … do not go by oral tradition, by lineage of teaching, by hearsay, by a collection of texts, by logic, by inferential reasoning, by reasoned cogitation, by the acceptance of a view after pondering it, by the seeming competence of a speaker … But when you know for yourselves, 'These things are wholesome … these things are praised by the wise; these things, if undertaken and practiced, lead to welfare and happiness', then you should engage in them." [6] Resonating with this is: "Better is one's own dharma, though imperfect, than the dharma of another well performed," quoted by Sherry Ruth Anderson, Ridhwan teacher - Ridhwan referring to a contemporary wisdom school, in which I'm a student. This dharma encouragement has been written even before the Buddha's time, in the Bhagavad Gita, the old Hindu text from around 3,000 BCE.[7] So don't be too much impressed "by the seeming competence of a speaker"; these are wonderful broad and generous invitations.

Potential for awakening

A few central notions in Buddhism, specifically in connection with Settling the mind in its natural state practice, are mentioned. In Buddhism, the basic roots of suffering and unease are seen to reside in the mind, and specifically: in spiritual ignorance, rather than in what is called "sin." The emphasis is on our fundamental true nature, our potential, and tendency to care. The mind is seen as containing many unskillful tendencies with deep roots. Underneath these roots it is bright and pure: this represents our potential for awakening. These notions are true for all early and later schools in Buddhism. The practices we did during the Shamatha Project, and that are described in this book, all contribute to observing, purifying and clearing of the mind, so that it can be open, spacious, full and free. The Buddha offered his teachings to everyone, in ways that they could be understood by any audience, and these teachings gave to everyone equally the possibility of achieving liberation through personal effort. Regarding dust and trust one might say: the less dust, the less effort. More dust, for a start, requires some more effort. The Buddha Shakyamuni lived in the sixth century BCE. For his time, in a culture with clear stratifications and exclusive castes, his approach was revolutionary. This remains the case, also in our time, within cultures drenched in certain values of sin, chosenness and damnation, inclusion and exclusion.

A practical approach in giving some glimpse of Buddhist notions is to follow what are called the "turnings of the wheel of Dharma." It is said that they refer to presentations that the Buddha gave during different phases of his teachings.

First turning of the wheel of Dharma

In the first turning of the wheel, "The Four Noble Truths" are emphasized. With a more accurate naming, this regards "The Four True Realities for the Spiritually Ennobled," as Buddhist scholar Peter Harvey suggests. In a way these four realities contain all of Buddhism to its core. The first True Reality regards the truth of "the painful," or suffering, encompassing the various forms of "pain," gross or subtle, physical or mental, that we are all subject to, along with painful things that engender these. The second is about the origination of the painful, namely craving. The third is about the cessation of the painful by way of the cessation of craving; and the fourth regards the path that leads to this cessation. [8] Often, the terms suffering, unease, stress, and reactivity are used. An important notion is the interrelated nature of all phenomenal reality. This includes our ordinary confused minds. If we can understand, "see through" the complex relationships of our own emotional and conceptual patterns, the mental conditionings, and the behaviors instigated by them, then the cycle of suffering can be broken. Relevant for the explorations to be described, is the notion in Buddhism that at the root of suffering lies the idea of the existence of an inherently, independent, separate "self." Again: in Buddhism one is stimulated to not take any proposition at face value, including the propositions advanced in Buddhism. See for yourself!

Settling the mind in its natural state is one of the basic meditations that can contribute to the research of "self." Vipashyana insight practice takes one further on the path of investigating this idea of a separate-self, leading to a possible confirmation or falsification. The stated position – about the absence of a separate inherent self – certainly does not exclude the fact that, in daily language usage, we speak about "self," and that we all dispose of a combination of functional inner skills, that may feel like self. Right now, you are reading this, later you may go for a walk, it is you who is doing this. Still, you do this in a context of interrelated phenomena. The notion about a not inherently existing self can be left open or taken as a working hypothesis. Nothing needs to be disposed of: seek and find out for yourself. Padmasambhava, addressing "the illusory separate-self sense" was quoted in 2.5. For one more voice on this seemingly separate-self: Buddhist scholar Sarah Harding, in her clear exposition of the three turnings, states: "This is considered to be a false notion, since upon direct examination through meditation and analysis, such a self cannot be found." [9] Some felt sense of this interrelated nature of all phenomena I find in my diaries:

Diary, September 2007

> How interesting, that contributing in these measurements in some way underscores the living in interdependence, and feels like supporting the melting of feeling separate ... We contribute with the measurement data, gathered at a certain moment, representing the state of our body-minds at that moment, in a situation in which a number of variables is "controlled" from the outside. We have no direct control over our skin resistance, brain waves, immune factors. We are "looked through" in sort of a bare humanness, and we do not receive feedback in personalized data. We all contribute to the atmosphere in which the data are taken, knowing that this atmosphere is also influencing the research outcomes. In that way, interdependently, we "co-create" the data ... There is a responsibility, and a personal influence; at the same time, in my feeling, a separate attached "I" drops out of it!

The first turning, sometimes referred to as Shravakayana (Foundational Vehicle), is often connected with insights emphasized in early Buddhism. In early Buddhism Theravada (the "Teaching of the Elders") is an important school.

Second turning of the wheel of Dharma

The second turning brings forward notions connected with Mahayana ("Great Vehicle," mentioned as starting from around 100 BCE). The two great notions that were emphasized in this phase were emptiness and compassion. Not only the self but all phenomena are interdependent and empty of separate nature: the seeming separateness refers to constructs, conventions.

In both Theravada and Mahayana literature, shamatha practices take a role in the triad of ethics, mental balance and wisdom. Mental balance and ethics (including practices like the Four Qualities of the

Heart, like we did during project retreat) support each other, as do shamatha, calm-concentration, and vipashyana-insight practice.

Settling the mind in its natural state, while being a concentration practice, generates insights in what is named "relative" mind (the psyche, individual mind, including psychological phenomena) and in that way contributes to the research of self. However, for the realization of emptiness, and mind in absolute sense, formal insight practice is needed. Emptiness refers to potential, to openness, spaciousness. Emptiness addresses the interdependence of all phenomena in the sense that there is not a self, a flower, a planet that is 'separate,' that is not part of a greater matrix of interdependent phenomena. As Harding states, this truth is referred to as *absolute truth*. On a relative level, the interrelated existence of phenomena and the functioning of cause and effect, *karma*, are considered the *relative truth*. To be able to comprehend these two truths simultaneously is to maintain a "middle path," without falling into extreme notions. Compassion arises in the recognition of how we live in ignorance: compassion for ourselves and others. The ability for experiencing, expressing and "living" compassion can be facilitated and trained, as meditations in this book will show. Starting to understand the true nature of emptiness naturally invites compassion, promotes the experience of selflessness, and the wish to help, to liberate all sentient beings.

Third turning of the wheel of Dharma

The third turning brings in a further development in Mahayana, with the notion of *Buddha nature*, the inherent potential for enlightenment, now in a more elaborated way with the deep longing to liberate all beings. Next to the eradication of negative qualities, awakening increasingly includes the full manifestation of all positive qualities, our human potential. Vajrayana (the "Indestructible Vehicle") and the Tibetan Buddhist Essence traditions of Mahamudra and Dzogchen will be addressed in later chapters. Our practice Settling the mind in its natural state has a close connection with the Essence traditions, in being seen as the shamatha practice for Mahamudra and Dzogchen.

Fourth turning of the wheel of Dharma

There are various ways in which a fourth turning of the Wheel is referred to. Recent voices address with this naming the coming of Buddhism "to the West." Let's broaden this, in the context of multi-media interrelated cultures and societies, to all of the planet-world. Important themes here are the often named adaptability of Buddhism, and topics and challenges like world peace and emphasis on socially engaged Buddhism. Isn't all Buddhism engaged? Themes also include eco-systems, gender issues, and relations between retreat and living in cacophony society. Certainly there is the crucial theme of meditation and psychology. There is the need for integration of various manifestations of wisdom, the need, for every person on the path, for integration of psychological maturation and realization.

Meditation practice

In this global section with some pointers about Buddhism, not much attention is given to Buddhist meditation. Certainly, through all phases and "turnings," in all strands of Buddhism, the importance of a basis in concentration-calm, also referred to as staying or stopping, has been emphasized as crucial for increasing peace of mind. For peace of mind and inner freedom, shamatha-calm and vipashyana-insight meditation complement each other. This is a refrain song that you will hear throughout reading this book, sung in many tonalities. Underscoring this, next to Alan Wallace's plea: here is the voice of Thich Nhat Hanh, contemporary Zen teacher and actively engaged Buddhist. In his book *The Heart of the Buddha's Teaching* he addresses the two aspects as follows. In his view, we tend to stress the importance of vipashyana ("looking deeply") because it can bring us insight and liberate us from suffering and afflictions. However, the practice of shamatha ("stopping") is fundamental. If we cannot stop, we cannot have insight.

Let us now turn to the complement to the shamatha attention practices that we did during retreat: the shamatha Four Qualities of the Heart.

Chapter 3

Qualities of the Heart and *Tonglen*

In order to realize the most meaningful life you possibly can, envision, what would you love to offer to the world so that, as you are facing your own death, when death is imminent and you're looking back, you can do so with a sense of deep satisfaction ... (Meditation instructions, 3.2)

When reading about the shamatha attention practices, before my participation in the project, the thought came up for me: these are very interesting, clear, structured approaches, and I'm sure they will be very beneficial for my muddled mind. But where's the heart, the juice, aspiration, love?

The Four Qualities of the Heart are also referred to as The Four Immeasurables, or The Four *Brahmaviharas*, the last name meaning: the four divine states of dwelling. They are shamatha meditations, and more specifically: they are called discursive shamatha meditations. They are based in, and include, thinking, reasoning, contemplating, and imagining. Yet, this is not a dry sort of reasoning and analyzing. They are very much felt sense and they show the contrary of the habitual association of thinking and reasoning with dry mind. Anyhow, in many strands of Buddhism and Taoism, the notion for "mind" also includes "heart." As to Settling the mind practice, thoughts and emotions are included: this practice indeed regards "heart-mind."

The Four Qualities of the Heart meditations have been part of our program as an important complement to the non-discursive shamatha attention meditations. Wallace sees them as crucial auxiliary practices for the development of shamatha, being deeply meaningful in their own right.

In this chapter the Four Qualities of the Heart are described, together with an example for application, timely relevant during the

Project retreat. Meditation instructions are presented for Loving Kindness meditation, and for a combination of Equanimity with *Tonglen*. As Lucid Dreaming and Dream Yoga have been part of the program, also these practices are briefly addressed. For conclusion some additional "matters of the heart" are explored.

3.1 Four Heartful Qualities

The Four Qualities of the Heart, stemming from early Buddhism, are virtue practices. The way they are contextualized here is Buddhist, but of course these four qualities are recognized and cultivated to some degree in all great wisdom traditions. Roger Walsh, in his book *Essential spirituality*, gives a wonderful overview of seven central practices to awaken heart and mind, presenting examples from Taoism, Hinduism, Confucianism, Buddhism, Judaism, Islam and Christianity. To quote just some examples for one of these seven, regarding love: "Wherever you are, whatever your condition is, always try to be a lover" (Rumi, Sufi-Islam). When asked about benevolence, Confucius said "Love your fellow men" (Confucianism). And: "Grant that I may not so much seek … To be loved as to love. For it is in giving that we receive" (St Francis, Christianity).[1]

The Four Qualities of the Heart can be seen as "skillful means" in a sense that they supplement both calm and insight practices. The Buddha has said that both wisdom without compassion, and compassion without wisdom, are bondage. I trust that it may be added that calm without compassion is bondage, and compassion without calm is bondage as well. We need wisdom for not being overwhelmed when really becoming able to see the suffering in the world. If not approaching this truth in a way that we can cope with it ourselves, we won't be in a position to help others. And we need compassion when having cultivated calm, to not withdraw in (personal) bliss, or just use our enhanced faculties of directed concentration to be a sharper hacker on the worldwide web …

Alan presented descriptions of the Four Qualities of the Heart, of which we received the summaries on a handout. More elaboration can be found in his book *The Four Immeasurables: cultivating a boundless heart*. Next to a definition of the four states of awareness, he names their "proximate cause" or that which arouses this quality. Then, he refers to the "near enemy," or that which seems similar but is an unhealthy variant, and subsequently he names the "distant enemy," that which is diametrically opposed. After these, signs of success are mentioned, meaning that this particular quality has been a remedy against these enemies. In conclusion, remedies or support are named that can complement for lack of the quality in case, when this quality needs strengthening. What follows is a bit schematic, but I like the clarity of it.

Loving Kindness

This is the heartfelt yearning and vision for oneself and others to experience happiness and the causes of happiness.

Cause: seeing the loveable qualities of another person and oneself

Near enemy: self-centered attachment

Distant enemy: ill-will (intention to harm, and taking joy in another's misfortune)

Success: less ill-will, as well as an increase in friendliness and warm heartedness

Remedies: equanimity, compassion.

Compassion:

Compassion stands for the heartfelt yearning that all be free of suffering and its causes, asking: what can I do?

Cause: recognizing that others wish to be free of suffering; the helplessness of their suffering. You're poised to do something, or that something can be done

Near enemy: grief, sorrow and depression

Distant enemy: contempt and cruelty; taking pleasure in someone else's suffering

Success: our capacity for cruelty and delight in other people's misfortunes subsides

Remedies: empathetic joy, loving kindness.

Empathetic Joy:

This is the delight in other people's virtues, successes and joys.

Cause: attending to the virtues, successes, and joys of others and oneself

Near enemy: frivolous joy

Distant enemy: envy and cynicism

Success: reduction of envy and cynicism

Remedies: loving kindness, equanimity.

Equanimity:

Equanimity refers to even-heartedness, even-mindedness and impartiality. It means: attending without attachment or aversion.

Cause: taking responsibility for our own conduct and actions

Near enemy: cold and aloof indifference

Distant enemy: craving or attachment to those who are like us, and aversion to those who are different

Success: less craving, attachment and hostility

Remedies: compassion and empathetic joy.

One question that was addressed during the discussions was about the cause of equanimity, described as "taking responsibility for our own conduct and actions." Some wonderings about this I described in my diary.

Diary, September 2007

> It feels like a touch of paradox; which often signifies that something worthwhile is up. There is, on the one hand, the personalness of taking responsibility, standing for what you do and have done, recognizing, maybe apologizing for earlier actions. On the other hand there appears to be a "non-personalness" in the impartiality, even-heartedness, even-mindedness to others as self, and to self as others, without attachment or

aversion. No contradiction needs to be involved: when we are truly present with all that is, impartially, all-inclusive, the ego-personalness melts. There is a giving, receiving, doing out of non-doing in the sense of non-interference; of not creating an "us" and "them." Presence, awareness, naturally includes taking responsibility.

Of the Four Qualities, Loving Kindness, Compassion and Equanimity are seen as aspirations, while Empathetic Joy is seen as an emotion. All four involve certain states of consciousness, and with the practices these wholesome states can be cultivated and trained. Like the nine stages in attention shamatha, this is about the "development aspect" in state training. Practicing with the Four Qualities of the Heart during retreat made us more aware of our states through the days, overall; and when these were not so wholesome, we could practice one of the Four Qualities right on the spot. Loving Kindness would always be helpful and supportive. Feeling lonely, feeling bored, feeling irritated for having to wait in a long line before a meal, feeling an aversion to having to clean the toilets, suddenly feeling a pang of desire for strawberries ... all might come up. To some people we take more naturally, to others less. How does it feel to choose to intentionally sit next to a person that I feel less connected with? The person so often looking angry at me – in my perception –, the one who's always dragging his feet, the one whose socks smell bad. Compassion practice could be in place ... And of course, the feelings that come up relating to the retreat: am I practicing enough? Am I doing well enough? Will I be scoring well, for the measurements? Might I be one of "the better ones," or should I worry about lowering the averages, being a burden to the project? It is the attachment that may make such questions problematic. The last few questions address our feelings about ourselves and the presence or absence of self-esteem, in a psychological dimension. In the Four Qualities of the Heart practices, a more spacious dimension of our potential and beingness is invited.

3.2 Genuine happiness

For Buddhism there is a clear difference between fleeting pleasures and conventional happiness on the one hand, and genuine happiness on the other. Conventional happiness is stimulus-driven, meaning that it requires sensory input, for instance a pleasant thought, a nice view, a tasty bite, a sensual touch. When the stimulus is not there anymore, the pleasure recedes. Genuine happiness is not stimulus-driven. It is not to be found in agreeable stimuli, but in dispelling the inner causes of suffering and discontent. When we make the aspiration: may you be happy, may you be free of suffering, it touches me always when we add: may you know the sources or causes of happiness, may you know the causes of suffering. That is where the insight lies, and the source for change in one's way of life. Genuine happiness necessarily entails the pursuit of virtue and understanding. It connects with leading a meaningful life, with aspiration, and Loving Kindness for oneself and others.

While you read the instructions for the following practice, again feel free to join us in the shrine room.

Loving Kindness – guided meditation

Here is an example of instructions by Alan Wallace in our morning session. Loving Kindness: the time is 9 am. Location shrine room, after yoga, early meditation, breakfast, washing up and cleaning:

> Settle the body in its natural state ... settle the respiration in its natural rhythm ...
>
> Now, let us arouse the creative faculties of our awareness in the meditative cultivating of Loving Kindness, initially focusing on yourself. In so doing I invite you to bring forward your own *vision of your own flourishing,* the type of well-being, happiness, fulfillment and meaning that you would love to realize. Let your imagination soar and envision your own happiness. Allow this aspiration, this yearning for such happiness to arise, wishing yourself well. And consider the hypothesis that you have the

ability to envision such fulfillment because indeed there is a capacity to realize this ideal. In Buddhism we name this Buddha nature, primordial consciousness, pristine awareness, the dimension of our own awareness that is primordially pure: brightly shining mind. If you wish, symbolically imagine this pure dimension of awareness as a radiant white orb of light. With each out-breath, imagine rays of light from this orb sent off your chest, the heart chakra; emanating out, fulfilling, saturating your body and mind with this light of Loving Kindness. May I realize my highest ideals, my fulfillment, genuine happiness. In order to realize such fulfillment, various causes and conditions must come together. Some of these are from outside. We do need support, basic requisites of life, food, clothing, medicine. There are other more intangible requisites, conditions from those around, from the environment at large that are needed in order to realize this ideal.

What would you love to receive from the world around you, in order to realize your own deepest aspirations? It is quite clear that reality often does not rise up to meet our mundane desires; but maybe there is something in the very fabric of reality itself that does align itself with these reality based aspirations. Consider that this may be so. With each in-breath imagine drawing in from the world around you the blessings of the world and support for our own deepest yearnings. Imagine these coming from all directions, converging upon your body-mind, converging upon your heart. With each in-breath, arouse this aspiration of Loving Kindness. May I meet with all the causes and conditions from the environment that are needed. May I realize the fulfillment that I most deeply seek. Imagine it to be so.

In order to realize such fulfillment, clearly there must be transformation within, as well. *How would you love to transform,* as a human being? With what qualities would you love to be imbued, of which qualities would you love to be free? With each in-breath, imagine drawing in the light of blessings from the world. With each out-breath imagine rays of light emanating from your heart in realizing this transformation, this purification.

And finally ... In order to realize the most meaningful life you possibly can, envision, *what would you love to offer to the world,* so that, as you are facing your own death, when death is imminent and you're looking back, you can do so with a sense of deep satisfaction, a sense that this was a life well led. What would you love to offer to the world? With each out-breath imagine life flowing through your heart to the world around you, offering your best, Loving Kindness for all beings including yourself: may we all be well and happy. We are cultivating the causes of such genuine happiness.

And now, release. Let your awareness rest in its own nature. Let's bring the session to a close.

We practiced many variations on these "heart themes," for instance with others brought to mind. Loving Kindness can be done for a very dear one, like a very good friend, a partner, one's child. It can also be cultivated for a person that one knows more casually, like a person in a shop, or the bus driver you often see. The practice is more challenging with a person that one feels aversion toward. The invitation is to bring to mind this person's aspirations, and to bear in mind that this person's heart desires and aspirations are just as real and worthy as those of yourself and your loved ones. Which is not always easy ...

We can expand the scope of beings toward whom we practice Loving Kindness: the persons in the room, the house, the village, city or region. And then up to the country, the continent, the planet and beyond. And we can include all living beings. For us on retreat, that included the deer around us, the bears looking for something to eat in autumn, the last mosquito that bit me; animal species threatened with extinction. I love the practice: it feels very spacious and inclusive, toward others and myself, helping to let seemingly separating boundaries melt.

3.3 *Tonglen* and state terror

The Four Qualities of the Heart have often been practiced together with *Tonglen*, a later Mahayana practice. *Tonglen* is translated into: "giving and receiving," or giving and taking. It can be practiced in different ways, toward different sentient beings and situations. One visualizes taking on the suffering and pain of beings and sending them happiness and ease. It may be practiced, for instance, in a way of compassionately drawing in the darkness of suffering, and lovingly sending out the light of happiness and its causes.

Diary, September 2007

> However much we are secluded here, still the world comes in, which is good. While we are sitting in the shrine room, for the 5 o'clock exchanges, Alan sometimes shares some news, which he gets from the web. Monks and nuns rising up in Burma (Myanmar) against state terror. We hear that the military have shot 9 persons, and imprisoned hundreds …
>
> Today we meditate, The Four Qualities of the Heart and *Tonglen*, visualizing an orb of white light of pristine awareness at our hearts. Breathing in: may you, in Burma, be free of suffering and the causes of suffering; taking in the dark cloud that burns without a trace in the orb of white light. Breathing out: may you know peace, happiness, flourishing and the causes of happiness. This is sent both to those who suffer, and those who make others suffer. These last ones suffer themselves, often with fear. This aspect of sending to both of them: so important, addressing our duality-reactivity …
>
> Later, in discussion, to the question what we can do in such a situation, Alan responded in a way that touched me. He gave a sort of "binary" response:
>
> 1) go in: meditate, prepare the body-speech-mind for the right action, for the moment that the opportunity offers itself; and

> 2) go out: in activism. As he clarifies: "it's there in many tastes and ranges, from wholesome to unwholesome." Too often activism causes getting polarized and reified: right versus wrong, we versus them, leading to unwholesome situations. People also bring their afflictions in their activism. There are other possibilities. For instance, Bush said in a speech: "We, the people in the US are outraged!" Alan advances a different way: not getting polarized but seeing the larger picture, like The Dalai Lama, or Bishop Tutu do. They have responded very quickly: "Please find a peaceful solution." They have no real power or money. Still they are two of the most influential people in the world.

This has been a very moving session for me. I remember the moment of breathing out: may you know peace, happiness, and the causes of happiness. Wishing this, both for those who suffer, and those who make others suffer. I know this reflex of wanting to "punish" the perpetrator with excluding him or her while sending love ... while this colonel is most in need of it, frozen in rigidity, petrified in feeling threatened. He may be living in a stimulus-driven world of contraction, with the effect that no information can enter that might falsify his paranoid worldview. Indeed, if some sparkle of "knowing the causes of genuine happiness" could come in, this would inevitably lead to change in his behaviors. It would make a lot of difference for Burmese citizens.

Also I've known this concern, as I started practicing *Tonglen,* of worrying about "receiving" dark clouds of pain and suffering, and not being able to carry this burden. And then it became clear that we don't practice from habitual separate-self sense. There's this orb of white light of pristine awareness at our hearts, our true nature. Knowing this brings in a change of perspective. Suffering is transformed. *Tonglen* practice can never harm the practitioner; on the contrary. It supports us in getting more in touch with our true nature.

This is an important point: with a foundation of shamatha, in which we have become somewhat familiar with the process of letting go of attachment to thoughts, it is possible to practice *Tonglen* in a way of not letting the incoming clouds stick, trusting that what goes out is from a boundless well. This practice unleashes the power of our Buddha nature, and it is done from Buddha nature, not by separate-self sense.

3.4 Heart meditation integrations

Following is an example of the instructions that Alan Wallace gave for a meditation on equanimity and *Tonglen*, toward oneself and others, and toward different persons in one's life, in situations that will always change.

Equanimity and Tonglen – guided meditation

As you read this, allow yourself to meditate with us. Take up the themes for yourself. Focus your attention ...

> For the first three of the Four Immeasurables, cultivating these as equally as possible with respect to all sentient beings is absolutely integral to each of these practices individually. It is the case, because without that they never develop into "immeasurable." So, it's already built in. The fourth Immeasurable is almost like the "icing on the cake." It goes back to the issue of impartiality as a central theme, which then undergirds the other three. We will now integrate the practice of *Tonglen*, from the inter-Tibetan current, into the practice of equanimity.
>
> Let's settle the body in its natural state, the breath in its natural rhythm ... Let your awareness rest in its own nature.
>
> And now, if you will, symbolically imagine this dimension of your awareness, pristine awareness, as a pure radiant orbit of white light in the center of your chest, the heart chakra. Then, returning to this large space of awareness that is larger than and subsumes the smaller subspace of your body-mind: arouse your

imagination, your intelligence. Bring yourself to mind, specifically on *an occasion in your life when you were at your best.* A period, a day, an afternoon. You can truly delight: there I displayed the best of me. Bring this person to mind: you on that occasion, manifesting the potential for goodness, for virtue that is within you. With each out-breath, breathe out to this person (you) the Loving Kindness, the yearning: may you be well and happy. With each in-breath, imagine, as you arouse the heart of compassion, this person's faults, defects, mistakes, mental afflictions being drawn and dissolved in your heart, as you arouse this yearning. May you be free of suffering and the causes of suffering. At each out-breath: breathe out the light of Loving Kindness, with each in-breath, draw in the darkness of mental afflictions and resultant behaviors, and dissolve this as a dark cloud into the radiant light at your heart.

Shift now to another occasion in your own life, *just a normal day,* neither one with great ups nor downs. There is a constant. It is that each day of your life there is this on-going flow, this current of yearning to be free of suffering, find happiness. Attend to yourself on such a day, a normal day. Breathe out the light of Loving Kindness and breathe in the heart of compassion.

And now shift to a darker period in your life, it could be a day, a longer period, *where the worst of you manifested.* Your mental afflictions dominated your behavior in perhaps very regrettable ways, you lost your good sense, your sense of judgment. There may have been injury of yourself, and others as well. Even in the midst of the darkness of delusion in such an occasion, still there was this yearning, however misguided, misdirected, to be free of suffering, and find happiness. Tending closely to this person who you were on this occasion, with each in-breath, arouse this yearning: may you be free of suffering and the causes of suffering. With each out-breath, send out the light of Loving Kindness: may you find happiness, may you cultivate the causes of such happiness. And let these appearances in phases of your life dissolve back into the space of awareness; all of them having their reality, all of them transient.

Now, bring to mind vividly *a person with whom you feel great closeness and attachment,* in the sense that you feel this person's presence in your life is very important for your own well-being. If this person would vanish this would leave a gaping hole. It may be a spouse, a dear friend, a child, a parent. Attending closely to this person, attend also to what it is about your relationship with this person, and this person's qualities that arouses such a strong sense of attachment, why this person is so important to you. The Buddha declared: "All that comes together comes apart." This relationship has not always been present and it will on some occasion, sooner or later, change, and vanish. Causes and conditions that gave rise to it will shift, and something else will take its place. This is certain. In all the vicissitudes of our human interrelationships there are constants, and one constant is that this person, irrespective of his or her relationship to ourselves, always seeks happiness and wishes to be free of suffering. Each out-breath, breathe out this light of Loving Kindness, each in-breath, draw in the darkness of suffering and its causes, and dissolve it. May you like myself be well and happy, free of suffering and the causes of suffering. Let the appearance of this person dissolve back into the space of your awareness.

And now, vividly bring to mind *a person toward whom you don't feel a strong emotion, attachment or aversion;* perhaps someone you even know quite well. This person seems to assert little impact on your sense of well-being. This neutral relationship too came about as a result of causes and conditions, and too it will inevitably change. It has happened that such a person becomes a bitter enemy, and it has happened that such a person becomes our dearest friend, or becomes a total stranger. Again, the constant is this person's desire for happiness and freedom from suffering. Breathe in, breath out as before. This person is every bit as worthy as we are. Let this person too dissolve back into the space of awareness.

And now, bring to mind *a person to whom you feel aversion*. Perhaps this person has harmed you personally in the past, or maybe it is more this person's presence and general behavior. You may feel that the world would just be better off if this person vanished. Reflect on what it is about him or her that makes this person so disagreeable, that arouses such aversion. This relationship too has come about in dependence upon causes and conditions, with all of them in a state of flux. It was not always such a hostile relationship and it is certain that it will not last. This person could become a dear friend, or simply fade into the status of a stranger. Again there is the constant: this person, like each person here, always wishes for happiness and freedom from suffering, and is worthy to realize his aspiration. Seeing through the veneer of this person's behaviors, mental afflictions that temporarily dominate his mind – at least in our imagination –, arouse the heart of Loving Kindness and compassion with each out and in-breath. And for just a moment dissolve all appearances. And rest for a moment, with no objects ...

Let's bring the session to a close.

Gratitude

In case you are not feeling so familiar with imagining or "seeing" this orbit of white light at the center of your chest: you may know it, trust it to be there. When light and darkness and the visual sense are not so familiar here, you may include other senses. Darkness, for instance, may feel sticky, smelly.

In this meditation we practiced witnessing our own feelings in various circumstances where the best and the worst in us and some range in-between manifested, while cherishing impartiality and empathy. We combined this with *Tonglen*, giving and taking: like giving love and support, and taking in faults, mistakes, suffering. And we did the same toward others. With this we cultivated our understanding of the truth of impermanence.

In relation to qualities of the heart, it was often mentioned by Alan that really we are standing on the shoulders of so many, who inspired the evolving of this Shamatha Project. On the shoulders, for instance, of those who have made this possible in a financial way by sponsorship. And certainly also, the shoulders of the many contemplatives who paved the way during the past many centuries, who have cultivated their practices and transmitted these to subsequent generations, who have selflessly offered their abilities and energy. Many Tibetan contemplatives have gone through a lot of suffering, in the past 60 years. They are giving their lives for the Dharma. We are standing on their shoulders, in deep thankfulness and appreciation.

I have felt this profound sense of gratitude, these states of awareness that the Four Qualities open to; thinking of these contemplatives, in refugee camps, in diaspora, chased from their land and cultural home. Also, in my imagination I have been with the Burmese monks and nuns, just these months, who are suffering the consequences of their rising up. Rising up, maybe not so much against anybody, but for Truth.

In later chapters more guided Immeasurable meditations will be presented, including in relationship with Settling the mind in its natural state practice.

3.5 Extra: Lucid Dreaming and Dream Yoga

Next to teaching the shamatha attention and heart meditations as described, in the later part of the expedition Alan has offered teachings on Lucid Dreaming and Dream Yoga. These could be applied as an added insight practice, allowing those thus motivated for a 24 hours to 24 hours practice.

A non-lucid dream refers to dreaming at night, while you don't know that you are dreaming. When you recognize that you are dreaming while dreaming, you become lucid, and are now in a lucid dream. The aim of Buddhist insight practice is to become lucid at all times, in that way to wake up to all states of consciousness, both during day and night. Also here, calm practice is a great complement and support to this insight practice.

For Dream Yoga, the first step is to recognize that you are in a dream, that you become lucid. Then, that you are able to maintain the stability and vividness of that recognition. Special of dream yoga is that you start practicing to transform the contents of the dream: you may change yourself, and everything that objectively appears to you in the dream. Regarding nightmares, when you meet a horrific dragon, should you be afraid? Or could you say, like in the practice of Settling the mind in its natural state, that nothing can harm your mind, whether or not thoughts have ceased? And how about changing, within the nightmare-dream, the dragon into a little lizard? You can do it, your own presence in the dream and all other phenomena consist of illusory manifestations of consciousness. As long as you don't grasp at them, reify them, make a "thing" of them, more real and substantial than they actually are, you can't be hurt.

A lucid dream is a perfect laboratory for the first person study of the mind. With shamatha concentration-calm we develop a stable, vivid awareness. It's like a telescope launched into orbit beyond the distortions of the earth's atmosphere. This provides a platform for exploring the deep space of the mind, in vipashyana insight practice. And here, with the insight practices of lucid dreaming and dream yoga we get back at questions about the mind: what is the mind? What is inside, outside the mind? As to what might be in the mind, are these true representations of what we perceive "outside"? What is real? Alan gave us practices to do during the night, but also during the day. A regular check: "Am I dreaming now?" is a good start. According to Tenzin Wangyal Rinpoche, teacher in the Tibetan Bön tradition, normally the dream is thought to be unreal, as opposed to real waking life. But normal waking life is as unreal as a dream, and in exactly the same way. Dream yoga applies to all experiences, to the dreams of the day as well as the dreams of the night. The path is practical and suitable for everyone. Yet, while the use of dream yoga to benefit us in the relative world is good, this is a provisional use of dream, he emphasizes. Ultimately we want to use dream to really awaken, to liberate ourselves from all relative conditions, and not just improve them.

Well, I didn't really get deeply into dream yoga, as much as I would have loved! But I did get a taste for lucid dreaming.

Diary, October 2007

> Dream: I drive a car, there are road-works. I've been told to "fasten my seatbelt." Suddenly we are led through a narrow lane, with curves; two metal grooves for the wheels, with a deep pit underneath. One wheel slips off the narrow lane, and then the car, with me, is hanging over an abyss ... hanging, it seems, for some time. At first, anxiety. Then coming in more lucidity: this is a dream! Nothing really can happen. The anxiety shifts into a sense of quiet excitement. I stay in this position for a while. How special, just hanging in a car, tolerating the sense of un-balance that may any moment tip over. How about letting myself fall ... Then waking up.

Alan has given courses in these Tibetan Buddhist practices together with Stephen LaBerge, psychophysiologist and dream researcher. In his explanations, Alan addressed the view of LaBerge that dreaming can be seen as the special case of perception without the constraints of external sensory input. On the other hand: perception can be viewed as the special case of dreaming constrained by sensory input. So, the only real difference between waking and dreaming experiences is that waking experiences arise with sensory input and dreaming without it.

With shamatha we un-dust and polish the lenses, for optimal observation. In Settling the mind shamatha meditation we are practicing non-grasping, to become able, with a better telescope or microscope, to make finer differentiations. With vipashyana we get more insights into what reality is, while awakening to truth. With vipashyana we "see through," we see.[2]

In the Daily Experience Questionnaire we participants had to log all shamatha attention practices and the Four Qualities of the Heart. Next to that, we reported on our dreams, and possible effects of dream yoga. More dreams are described in Chapter 5.

3.6 More matters of the Heart

In what follows I address some additional themes that are "close to my heart."

Dust and trust, heartful explorations

"The less dust, the more trust," dust and trust ... Next to the few remarks addressing dust and trust in the Introduction, I would like to explore these notions somewhat more extensively. Trust has a lot to do with Heart, and The Four Qualities of the Heart: the virtue practices. Trust is formally recognized under morality, in the Buddhist triad of morality, meditation and wisdom.

As Daniel Brown, Mahamudra teacher and psychologist, describes: trust or faith has been called "the basis of all positive qualities," "the basis of sustained interest," and "the mother of all that is positive." Three types of trust are distinguished in Abhidharma Buddhist psychology sources: (1) trust that is belief, (2) trust that purifies, and (3) trust that desires direct experience. [3] Regarding the first: importantly, this is not blind trust – still, it requires openness, and a certain decisiveness. This first aspect of trust refers to the initial trust. Second: as this is about purifying, this sounds like trust that diminishes dust! And regarding the third: this is said to relate with a yearning for something. Placing this in a somewhat broader context: also, this seems to connect with the Buddha's proposition – as addressed in the Kalama Sutta – to investigate and experience for yourself. In that sense, it's about not just going with what you hear and read. It is about un-dusting the notions and positions you are familiar with, about tolerating falsification, about opening for new spaces, insights, for new more integrative knowledge. Also, it is about un-dusting the instruments that you investigate with! Well, in a special way this is what's done by both meditators and researchers.

To open ourselves to new perspectives and knowledge, also for the explorations in this book, we do need some degree of trust. We need some trust, not only in the perspectives offered and in those who present these, but in our experiencing, and in ourselves. In the end it is you, I, we, who decide the "truth or falsity" of any theory or set of understandings for us, by deciding whether we can trust our experience, our instruments, our measurements. As psychotherapist Elliot Ingersoll rightly reminds us: "If we decide we can't trust our experience, can we trust ourselves to make the decision about whether or not we can trust our experience?" [4] Next to the three types traditionally described in Buddhism, we may add this basic type of trust: this trust in our own senses, our experiences, with a certain degree of basic trust in ourselves.

Focusing now on the explorations as described in these chapters: how can we support the process of letting dust in our eyes dissolve? This question may be seen as a leading thread in my journey with the meditation practice Settling the mind in its natural state and this book. Philosopher Ken Wilber, who is a longtime meditator, outlines three basic steps in the process toward what he names "good knowledge."

First is to decide what you want to find out, and then to engage in "practices" that address this area. So, this is about the instruction "if you want to know this, do that," *the experiment*, the enaction. Let's say, for me the question is: "does Settling the mind in its natural state practice support the dissolving of dust, and the increase of trust in my own discernment, in myself?" Well, then clearly, do the practice, and explore it!

Second is the *experience* you have as a result of following the instruction, as named in step One. It is the awareness, the data that comes out: an illumination of the phenomena brought forth or enacted with performing step One. In this book: this is mostly about my experiences in doing the meditation practices.

Third is a "*communal confirmation or rejection*," referring to the checking with others who have completed step One and Two. [5] During the project, in the afternoon sessions, we brought in what had been our findings. This then could just be shared in the interpersonal sharing field. Additionally, others might add their experiences – all experiences having their own validity. We might find sequences, see patterns. With the guidance of Alan, in the group and in our individual sessions, we might discover that some experiences were more helpful than others, on the path. Experiences might then be understood, indeed, in an interesting amalgam of Buddhist contemplative and Western scientific contexts of giving meaning. Regarding Settling the mind practice: in this book I describe what others have told and written about this practice, what information is available. This step of communal confirmation is crucial: while it is important to trust ourselves, this also requires that we get to see the many ways that we are able to deceive ourselves. Our conventional mental functioning includes many ways of distorting perceptions, sensations, and emotions.

At the end of this book-journey, when having taken these three steps, when I'm qualified to respond to the question "Does shamatha meditation, and specifically Settling the mind in its natural state practice support the dissolving of dust, and the increase of trust?" I'll come back to these points.

Right from the Heart: four working hypotheses

In Chapter 1, research questions for the Shamatha Project have been presented, together with the ways that the researchers have addressed these questions.

When I was selected for participation in the Project, I took up courage and shared that I would love to do a little structured exploration also for myself and my studies. This felt like a modest psycho parallel process, and a personal adventure within the large Shamatha Project research. Next to the project-diaries we were requested to produce I made some extra personal notes; for which there was space in the Project. In the beginning I had no idea where this might be leading. I just brought fresh notebooks and pens for jotting down, reporting, as mentioned. During the first month of expedition, a number of questions came up for me, which evolved into, what I will call "working hypotheses." At first, this was in a rather coarse fashion. Gradually, also in my musings after the expedition, they have evolved into what is presented here. They have helped me in providing a matrix for structuring the large amount of experiences, impressions, questions and tentative understandings that emerged from my curiosity and on-going heartful engagement with the Project. Well, "working hypotheses" has a somewhat solemn sound ... the terminology may be a bit tongue-in-cheek, but the content of what it's about to me is very serious.

I named four hypotheses. I will come back to these hypotheses in Chapter 11, and present my findings to the themes mentioned.

Hypothesis 1: "Shamatha practices, and specifically the meditation practice Settling the mind in its natural state, provide first person phenomenological understanding of dynamics in the psyche."

I thought that I knew a bit about the psyche (that is the personal mind), from outside and inside – and I learned a lot more. The interesting thing is that the mind, while learning about itself, keeps transforming.

Hypothesis 2: "Shamatha practices, and specifically Settling the mind in its natural state, in a relational-compassionate context, can have wholesome effects for any human being, irrespective of worldview. This is the case because, also when stripped of explicitly Buddhist context, this practice can be presented as an attention and emotion regulation strategy and training in its own right. As the practitioner progresses in the training, worldview-related thoughts will less and less be grasped at and identified with."

So, this second hypothesis is about allowing the mind's working to be more effective and wholesome: this regards attention and emotion, with generalization into everyday life (including the everyday life effect for us participants during these three months).

Hypothesis 3: "The meditation practice Settling the mind in its natural state invites for inner turbulence. At the same time, it presents empowering ways of making sense of attentional and emotional turbulence experiences and challenges. Moreover, the practice offers tools for coping with these challenges through enhancing the practitioner's capacity to focus and refine attention, to not be distracted by thoughts and emotions that come up, and by enhancing the capacity to not grasp at and identify with upcoming thoughts and emotions. These are tools that nourish transformation in a wholesome direction. This meditation may be adapted for various applications that will relieve suffering."

Also this hypothesis, regarding coping with attentional and emotional turbulence and challenge, has certainly been inspired by my own experiences, as described later.

Hypothesis 4: "Shamatha practices, and specifically Settling the mind in its natural state, offer ways of directly experiencing, and becoming more familiar with various dimensions of psyche and consciousness, with gradual refinement in perception. This creates perspectives for collaboration in consciousness research between researchers in various fields, like for scientists and contemplatives."

The fourth hypothesis addresses the understanding of consciousness, getting more familiar with exceptional states of consciousness, and collaboration in research. As for the Project, these aspects regard the scientists and me, 1. with myself as a curious guinea pig for the Project scientists, and 2. as a contemplating researcher and inquiring meditator for the participant observer in me.

It may be noted that the named hypotheses include few direct references to Buddhism in general, Buddhism as it evolved in Tibet, or Buddhist psychology. This, by itself, may be viewed as an expression of the deep wisdom and insights in the human condition that Buddhism offers. At the same time, the description and contextualizing of the meditation practice of Settling the mind in its natural state in the Tibetan Buddhist tradition, in chapters to follow, will underscore the richness of the sources that nourish the practice.

Heart for integrative views

Here's another approach that is close to my heart. Connecting with the fourth hypothesis about collaboration of scientists and contemplatives: it may be helpful, for understanding and honoring these objective and subjective and more various views, to present at this place a number of fundamental perspectives that weave through this book. These perspectives are based in Integral Theory and the Integral model, as offered by Ken Wilber. This model provides a number of useful distinctions that relate to the multiple perspectives presented in these chapters. Why? The reason is that I find them so clarifying, they touch my heart, they help me to better see things in context. The core, to say it briefly: any phenomenon – including the Shamatha Project, the meditation Settling the mind in its natural state, heart-mind, you, and me – can be explored from four main integral perspectives. [6] Here I elaborate on the brief description of first, second and third person approaches, given in Chapter 1.

(1) The first perspective addresses how I personally, subjectively, experience an event, see and feel about an event: an individual interior view. This is the view from self and consciousness, from meditation and psychology. This perspective of self and consciousness, for short, is referred to as "Experiences." With a first person approach I use the pronouns like I, me, mine.

For the Shamatha Project, and throughout this book, this subjective Experiences dimension has clearly been given much attention. Regarding the meditation practices I describe how I have experienced engaging in them. The Experiences dimension also stands out in the diaries, with descriptions of what I went through, that day.

(2) The second perspective includes, next to me, how others see and feel about the event: this is about the intersubjective view from

culture and worldview; a collective interior view. This perspective is referred to as "Cultures." The second person approach refers to the person with whom I'm interacting, who's interacting with me: you, yours. And, when being aware of our sharing in this: we, us.

In the Shamatha project, as to the "we" components: there was not only everyone's subjective, but also our collective intersubjective experience of the expedition. We shared a motivation to participate, and some values and views. We were embedded mostly in, on the one hand, contemplative, and on the other, neuroscientific cultural terms. Generally, some meditation practices include more of a Cultures aspect than others. The Four Qualities of the Heart very much include the "we" view, the view of our common humanity. With these Heart meditations we practice widening the scope, with including not only friends but also those we are not very fond of. Also we can widen the scope in a geographical sense with persons all over the world, and with including all sentient beings. We can practice up to all-inclusive, really immeasurable, boundless states.

(3) The third perspective regards looking at an event in the sense of an exterior view, it is for instance – regarding our individual being – about the brain and neurophysiological measures. It may be named "Behaviors." This is the third person objective view of phenomena, individual exterior.

In the project, the perspective of neuroscience stood out: this is, as we saw, about objective measurements, material components, about his, her, my brain, about EEG patterns, assessments in blood, and saliva. On the basis of abstractions from multiple data bases, findings are presented. The event that is investigated here, offers information in the form of, you might say, an "it."

(4) Also a description can be given from the perspective of social system and environment, which can be referred to as "Social systems." This is about inter-objectivity, collective exterior view; the events as embedded in social systems and environment, "its," third person plural.

For the project, it refers to the exterior forms and behaviors of the project organization, and our group: living as a temporary close community, located in Shambhala Mountain Center, with a certain division of labor, a schedule to follow, roles and functions.

Four quadrants

An individual has direct access to experiential, behavioral, cultural and social-systems aspects of reality; you might say, this is what the individual is looking from. These are dimensions of her or his embodied existence. These four dimensions you may imagine, in a spatial sense, in a model with four quadrants: Experiences, Upper Left; Cultures, Lower Left; Behaviors, Upper Right; and Social systems, Lower Right.

When looking at a specific phenomenon, the four connected perspectives enable us to look at it from many sides. Now, imagine me practicing the Four Qualities of the Heart, let's say Loving Kindness meditation. Look at this person practicing from four perspectives, evaluating the situation with four kinds of expertise. With the Experiences perspective: there is this person, feeling, thinking, imagining another person in the sense of "may you be happy." You can imagine this coinciding with certain EEG patterns, neurotransmitter dynamics, DNA and so on, in the meditator, relating to the Behaviors perspective. While practicing Loving Kindness, you can figure that this person is a member of a group, with certain cultural and ethical viewpoints, relating to the Cultures perspective. And there is a Social system, an organization that facilitates the sitting practicing of this person in a hopefully quiet environment.

In the quadrant view, a developmental aspect toward increasing depth and complexity is included, to which we will come.

Now that I have addressed the scientific context (Chapter 1, third person measurements, mainly objective) and the specific practices in the project (Chapter 2 and 3, mainly first and second person, subjective, intersubjective), we will consider a wider Buddhist context for understanding the practices, in the next chapter.

Chapter 4

Shamatha in Buddhism as it developed in Tibet

A colorful picture representing the nine stages in shamatha has been attached on the door to the shrine room ... When entering, there is a recognizing myself, and feeling recognized in struggling with elephant and monkey. (Diary, 4.2)

The representation of the Elephant Path, with its winding road and nine elephants, can be seen as one of the early comic strips or cartoons in history. I remember, as I was struggling at the beginning of the path, the feeling was not very comical. In Chapter 2 reference has been made to the nine stages, and in this chapter a more elaborate description of these stages is presented, together with an illustration of the comic heroes. One consolation is that, wherever we are in the practice, we are always represented: we're always in the picture, even when it's far off in the lower right corner, with a dull elephant and restless monkey-mind.

After the description of the blood and bones of the meditations in the way we were taught during the Shamatha Project expedition, this chapter takes a look with a bird-eye's view, and addresses some aspects of shamatha meditation as it is contextualized and practiced in Northern Buddhism. Attention is given to the relationship of shamatha and vipashyana, and to the nine stages and attaining substrate consciousness. Also, mindfulness and introspection and their various meanings are addressed, now in a more theoretical context. More contemporary perspectives on these notions are included. While I received part of this information in Dharma teachings by Alan Wallace during expedition, the details came to me, mostly, after the three-month expedition, from later understandings and readings.

4.1 Exploring shamatha meditations

You may have noticed that the term *shamatha* has been used in various ways, which can generate some confusion. For example: the use of the term includes: 1. a particular state in which one can allegedly focus on an object for an unlimited period of time (like in the expression: achieving shamatha), 2. a style of practice aimed at attaining that state (like in "shamatha and vipashyana," or the Shamatha Project); and 3. the aspect of any meditative state that constitutes its maximal stability. [1] This underscores the need to be clear about what is meant, in a particular context.

Northern Buddhism, Essence, and energy dynamics

A distinction is sometimes made between Southern, Northern and Eastern Buddhism. Shamatha is practiced in all regions. Southern is where Early Buddhism, including the Theravada school, is found, along with some elements that were integrated from the Mahayana. Northern Buddhism is the heir of the late Indian Buddhism, where the Vajrayana version of the Mahayana has developed; and Eastern Buddhism is where the Chinese transmission of Mahayana is found. Northern Buddhism connects with Buddhism in countries including Tibet, Mongolia, Bhutan, and Nepal. Buddhism may have entered Tibet in the fifth century CE, and it became established from the seventh century on. [2]

Within Buddhism, a distinction is made between the Sutra and Tantra methods for spiritual development. The term Sutras refers to what are named the discourses of the Buddha, connected with Early Buddhism. They include the Three turnings of the wheel of Dharma, referred to in Chapter 2.6.

As Daniel Brown explains: the Sutra methods, generally drawn from the practices of early Buddhism, are typically threefold: first, the preliminary practices, making body and mind fit for formal meditation, and second, concentration training to stabilize the mind. Third are the special insight practices to realize the mind's essential nature. Regarding Tantra methods: these broadly include two kinds.

First are the generation (or creation) stage practices, which entail complex visualizations for stabilizing the mind and manifesting its essential nature. Second, there are the completion stage practices that enhance these realizations at the subtle and very subtle levels of mind. These may involve manipulating the body's energy streams in order to make the very subtle mind more accessible. Reference is made to a third category beside Sutra and Tantra methods: the Essence traditions, that are exemplified by the Mahamudra and Dzogchen traditions. [3] Sometimes these are included under vipashyana, as insight in a wider sense. The Essence traditions are often seen as the living teaching of the Buddha, the living speech and heart of all enlightened beings. They involve relaxing our minds in the natural state, the original primordial state. From this vantage point everything is seen to be complete and pure as it is, right from the beginning. Settling the mind in its natural state is seen as the shamatha practice for these Essence traditions, where directly working with the mind takes an important place.

In the Essence context, Maitripa, an eleventh century Mahamudra teacher, who played an important role in bringing the Mahamudra teachings from north-eastern India to Tibet, describes three types of quiescence, or shamatha meditation. They are: (1) quiescence that depends on signs, (2) quiescence focused on conceptualization, and (3) quiescence that is settled in nonconceptualization. We may recognize in this triad the meditations that Alan Wallace has taught us: first, with a sign, the meditation on the breath, and third, awareness of awareness. The second, quiescence focused on conceptualization, is also named Settling the mind in its natural state. Maitripa gives a clear exposition. He describes the excessive proliferation of conceptualization, including such afflictions as the five poisons, thoughts that revolve in duality, thoughts such as those of the Six Perfections. He admonishes: " ... whatever virtuous and non-virtuous thoughts arise – steadily and nonconceptually observe their nature. By doing so they are calmed in nongrasping." The Five Poisons are: attachment, aversion, ignorance, jealousy, and pride. With the Six Perfections, reference is made to generosity, virtue, patience, vigor, meditation, and wisdom. Maitripa adds that awareness vividly arises clear and empty, with no object of grasping; and then is sustained in the nature of self-liberation, in which it recognizes itself. And he continues: "Again, direct the mind to whatever thoughts arise, and without acceptance or rejection, you will recognize your own nature. Thus implement the practical instructions on transforming ideation into the path." [4] Interestingly, this passage is from a chapter on the cultivation of quiescence. The last part suggests how this practice may continue into a path to awakening. More on Settling the mind in its natural state practice in the Essence context will follow in Chapter 7.

Subtle energies

Before briefly going into the relationship of shamatha and vipashyana, in Sutra context, I want to devote a few words to energy dynamics in Tibetan Buddhism. In mentioning Tantra, reference was made to the cultivation of the subtle energies of the psycho-physical body. Energy awareness, contraction and relaxation have been important aspects in the practice for me. Anne Carolyn Klein, Buddhist scholar and Tibetan meditation teacher, mentions how energy can be seen as that to which the Tibetan tantric and medical traditions refer as the "inner winds," *rlung* in Tibetan, also referred to as "lung." This corresponds in virtually all ways with what Indian Hindu as well as Indian Buddhist traditions refer to as *prana* and the Chinese as *qi*. The "inner winds" travel in the internal channels, or *nadi* of the subtle body. Reference to this energy has also been made in Chapter 2.5, in mentioning meditating for 24 minutes, one *ghatika*, the time of one cycle of energy through the body. All perception, and all mental functioning, any kind of cognitive or emotional processing, including what the West calls psychological, is always accompanied by some movement of the inner currents of energy, in these views. And, in relation to shamatha: also the most relatively simple Buddhist practice includes awareness of body and energies.

In the words of Anne Klein: "To sit and observe one's breath, to take the most classic example, is also to calm the breath, the body, and all the subtle currents moving through it." [5]

In case you have taken up the practices that have been presented in the preceding chapters, especially the breath meditations, you may reflect on these remarks about energy, with these questions: did you get a felt sense of these energy currents? And: is there a difference in your felt-sense, energetically, when you focus on the tactile sensations of the breath in the whole body, as compared with the abdomen, and the apertures of the nostrils? You may try it out, what is your experience?

Shamatha and vipashyana in Northern Buddhism

In Northern Buddhism calm and insight meditation and their relationship, as they had globally been in Early Buddhism, gradually became modified by the Mahayana framework of belief and motivation. What I'm describing as the understandings I received from teachers through the years in the sense of shamatha, vipashyana and Mahamudra, is mainly in Sutra context. However, they also include Tantra and Essence elements. The classical practices of calm and insight in Northern Buddhism find their basis in the texts by the eighth-century Indian teacher Kamalasila, on the stages of meditation. In the Tibetan context, these received thorough formulation by Tsongkhapa, founder of the Gelug school, who lived around 1400 CE. [6]

At the beginning phases on the path, in order to develop a good basis for meditation, and as a remedy for an agitated mind, four contemplations are recommended. These are referred to as the "The Four Thoughts that turn the Mind to the Dharma," or the four reminders. Alan Wallace taught them to us. The first regards the preciousness of the human life with leisure and endowment: this connects with the freedom and opportunities that a human birth gives us. Second is impermanence and the inevitability of death. Third refers to the suffering in samsara, the cycle of existence.

Fourth and last is: the nature of karma, the workings of cause and effect. With these four, an attempt is made to counteract certain basic, habitual delusions that we all carry around. For instance, the awareness of the preciousness of human birth may act as an antidote to our tendency to take our life for granted (and for me, this made me so aware of just temporarily functioning as, while not really living the life of a scientific guinea pig). The awareness of impermanence counteracts our disregard of our mortality. These Four Thoughts that turn the mind, as well as meditating on the Four True Realities for the Spiritually Ennobled (addressed in Chapter 2.6) are crucial as preliminary practices for purifying the mind. Also, these contemplations invite for the arising of the spirit of awakening, *bodhicitta*. Together with the named contemplations, it is recommended to practice meditations concerning the needs of others that inspire the Mahayana motivation of saving others. This is done by developing the Four Qualities of the Heart, as we have seen: Loving Kindness, Compassion, Empathetic Joy and Equanimity. Calm, insight, and heart go together, in this way.

According to Lama Sopa Rinpoche, contemporary meditation teacher, the immediate aim of Buddhist meditative practice is the perfect union of calm and insight, in *shamatha vipashyana yuganaddha*. Sopa Rinpoche agrees with Wallace and many others that one of the most important ways of explaining the shamatha path is from the point of view of nine stages of concentration, as referred to in Chapter 2.5, in a practical way. Below I give some more elaboration.

4.2 Shamatha practice and the nine stages on the Elephant Path

The representation of the nine stages of meditative concentration, with a succession of images of an elephant, a monkey and a monk or nun, can be found in many Tibetan Buddhist temples. Here we will look at them in more detail, from various sides: including practice, theory, and from lively picture and metaphor.

Each stage represents, compared to the one before, increasingly better focus, continuity, stability and vividness in attention. Different descriptions are used and different names are given. At this place I mention the terms used by Sopa Rinpoche and Wallace for the stages, followed by some elaborations that they make.

1. Interiorization, or Directed attention
2. Duration fixation, or Continuous attention
3. Refixation, or Resurgent attention
4. Close fixation, or Close attention
5. The disciplined, or Tamed attention
6. The pacified, or Pacified attention
7. The completely pacified, or Fully pacified attention
8. The one-pointed, or Single pointed attention
9. Even fixation, or Attentional balance.[7]

Attention, tamed and pacified

Chapter 2.5 has presented some aspects of these stages, in the ways they came across in our exchanges in the late afternoons during retreat. I didn't read too much about them then, as I thought: let me not be influenced by these descriptions, let me not bias myself too much. Then, exploring what had been written about the stages, after these three months, there certainly has been so much recognition.

Progression and attainment in these stages are described as happening by way of learning instructions, thinking, mindfulness, introspection, enthusiasm and familiarity. All these, more or less, seem to include the aspect of monitoring the practice and the quality of the practice: the feedback "looping" process that nourishes progress on the shamatha path. Introspection is noted specifically as the active agent in stages 5 and 6, Tamed Attention and Pacified Attention. In stage 5 there is, on the one hand, the danger of lethargy; and, on the other hand, as a consequence of too much heightening of the mind, the opposite danger of the mind's becoming over-stimulated or excited. This is increasingly brought under control in stage 6, where the mind is "pacified." Taming, pacifying, fully pacifying ... a lot of work needs to be done here, it seems! Stage 5, and certainly 6, in spite of their names, may be emotionally rough, as Wallace notes in *The Attention Revolution,* in describing the problems that persist, respectively: "Some resistance to samadhi," and "Desire, depression, lethargy, and drowsiness." I think I can testify to that to some degree, as will become clear in later chapters! It is in stage 7 and 8 specifically that enthusiasm is present, the practice itself now brings joy. For the ninth stage Sopa Rinpoche and Wallace describe the sense of serenity and felicity of mind and body that arises, when the body-mind has become so completely acclimatized to meditative concentration that all traces of physical sluggishness and mental uneasiness have been eliminated.

As Alan Wallace emphasized: these stages are not meant for measuring. Hopefully they can be helpful descriptive markers for recognition on the path. So, let's have compassion for ourselves, with our high expectations that are destined to fail. I'm reminded of Jason Siff, meditation teacher, who authored a book with a saying title *Unlearning meditation – what to do when the instructions get in the way.* Next to formal hierarchies, he addresses the hierarchies of meditation states that we tend to create for ourselves, with at the top what we consider optimal states of mind like mental clarity, mindfulness, equanimity, love, and near the bottom those that we consider gross or mundane. It's often implicit, and also it may be there in the meditation session as part of the running commentary on our experiences. Let's be kind to ourselves, and let's have some trust in the process. It doesn't work, beating ourselves up. Interestingly, with Settling the mind practice, these pressing high ideals, and the ruminations may be part of the grist for the mill of our meditations! They are observed, while grasping is released. Often, I think, it is not meditation that needs to be unlearned, but grasping that needs deconditioning! Maybe even some inner therapeutic processing is happening, by way of non-grasping to high ideals and self-judgment patterns, in this practice ...

A pictorial illustration of the Elephant Path

Following, now, is a pictorial representation of the nine stages: I got to know it well, during the three retreat months. I love this image, giving such a lively touch to the theories!

Diary, September 2007

> A colorful picture representing the nine stages in shamatha has been attached on the door to the shrine-room. I see the winding road, and the monk or nun starting the trajectory in the lower right corner. He or she variously looks desperate, eagerly motivated, serene, in my idea; I may project this into the picture ... When entering, there is a recognizing myself, and feeling recognized in struggling with elephant and monkey. And this curious little rabbit, briefly coming in, smiling, has some odd sort of an uplifting effect.

The picture is found on many *thankas*, and in many temples. In the following brief description of the nine stages illustration I connect with the explication by Lobsang Lalungpa, in his introduction to the classic *Mahamudra, the Moonlight* by Dakpo Tashi Namgyal.[8]

In Figure 1 you see a temple, in the lower corner on the right of the picture, indicating the need for serene solitude while practicing shamatha meditation. The monk (or nun? *She* is not included in Lalungpa's text, in many Buddhist texts ...) represents the meditator, and the winding path signifies the nine stages of meditative concentration-calm.

The first and lowest stage 1 is symbolized by the monk following an elephant and a monkey. At this stage the meditator finds it hard to concentrate on a chosen object, and discovers that his inner world is overcome by the restless senses, thoughts, and dullness. The grey monkey represents the restless mind, the grey elephant the heavy dullness. The rope and axe with the monk indicate the strong need for mindfulness and self-control in the early stages. The flame along

Figure 1: The Elephant Path in shamatha practice, a pictorial impression

the first six stages is representing vigilance, or introspection. As we saw: mindfulness watches over the mental focus of concentration, while vigilance, introspection specifically detects any emergence of distraction, in either coarse or subtle forms. Stage 2 shows some progress in concentration, as indicated by the white patch on both the monkey and the elephant. Stages 3 and 4 show the monk making more progress by clearing the restlessness and dullness that alternately interfere with his concentration. The monk is roping the animals, but the appearance of a rabbit on the elephant's back indicates the emergence of a subtle dullness. With gradually growing stabilization, the concentrated tranquility also produces a tricky inner ease and ecstasy that may rob the mindfulness of its sharp focus. In the illustration, stages 5, 6, and 7 show the monk gradually achieving greater tranquility, as visualized in the animals' widening white spots. In stage 8 there is near-perfect tranquility, as symbolized by the virtually white elephant and monkey. In stage 9 the monk is shown in solitude, absorbed in Attentional balance, with the equally restful elephant. This tranquility represents the serenity associated with what is called the "human plane of desire." Mastery of Attentional balance will lead to the next stages, the super-worldly and transcendental states. These are even higher states of absorption than we are addressing in this nine-stage context. They are referred to as the *jhana*s in Pali, *dhyana*s in Sanskrit.

The top of the image shows two monks riding on white elephants on two bands of multi-colored rays of light, representing the higher contemplative super-worldly states. One band, with its four states, corresponds to the natural consciousness on the refined plane of form, with the four absorptions. The other four states correspond to the formless plane. The sword in the monk's hand stands for the sword of ultimate knowledge that cuts through self-delusion. As I get to understand: this is the monk that returns from the higher refined form and formless absorptions path, a path that cannot lead to awakening. The sword of ultimate knowledge stands for the need for insight.

Disentangling from mind – guided meditation

The following meditation guidance, presented by Wallace in the first weeks of the project-retreat, connects with the first stages on the path, with overpowering elephant-monkey activity. It may be taken up when one tries to practice Settling the mind in its natural state while the mind is extremely full, busy, buzzing ...

On occasion you may find yourself so overwhelmed by the contents of the mind, obsessive thinking, emotion and desire, that you find that as much as you might like to practice Settling the mind in its natural state, you can't. Because its grip seems quite strong, and you can't make your finger straight, you can't disentangle yourself from it. Then two options spring to mind:

1. One is to *withdraw,* withdraw from the mind. For example, you may come to the breathing. All those thoughts, who can release them? I'm just going to focus on something that doesn't arouse any craving or hostility or any other mental affliction at all, just the tactile sensations of the breath that are so neutral. So get your mind off. As if you must take one step back and say, well, never mind, I'm going to withdraw from the problem.

Or, if you find the mind is too turbulent for that, then take another step back, do something like recite a mantra, or in the Christian tradition the Centering prayer, something to really get your mind off from where your mind is in the grip.

Or, when that doesn't do, get off your cushion. Go for a walk, look at the trees, read something, occupy your mind with something very different.

2. But, there's another approach: that is to cut right through the mind, to *advance.*

Here's this great morass of the hordes, barbarian hordes of the mental afflictions assaulting you. And now instead of cringing or waging battle with them, or observing them, crumbling with fear as they assault you, just make your mind like a blade, go right through and just be aware of awareness. Cut through your mind

> to that from which the mind emerges. Cut through the space of awareness that is free of the coloration of desires and emotions, free of the structures of thought, memories, personal history. Cut right on through the other side, and then you flank them. OK, try it. This practice of shamatha without a sign is not so much a matter of developing stability and vividness, but rather of unveiling or discovering the innate stillness and luminosity in the very nature of awareness itself, and we unveil it by releasing all grasping to subjects and objects ...
>
> Let's bring the session to a close.

Certainly, sometimes it was the right thing to do: get off the cushion, and go for a walk, opening up to the vast space of the sky, of awareness.

Diary, September 2007

> This afternoon I climbed up to Marpa Point, this high point of one of the hilly mountains on the terrain of the center, now a Buddhist devotional spot, with flags and piled up stones. It does feel like a sacred spot. It has been so for the Native Americans who lived around here before, and now is for Tibetan Buddhists. I'm wondering: how did the former inhabitants live around this spot? Did they have ceremonies here, perform spiritual practices, were they praying, singing, dancing ... do I hear distant drumming sounds? There is stillness, wide perspectives, presence, nature as it is. I was sitting on a large stone for a while, looking up and far, dissolving into the vast expanse. Until I got too cold, with freezing fingers. Not having gloves with me ... who expects to need gloves in September.

4.3 Attaining substrate consciousness and achieving shamatha

The accomplishment of the ninth stage of approximate tranquility, or attaining shamatha, is also referred to, in Theravada Buddhism, as "access concentration." Buddhaghosa, a fifth-century Indian monk-scholar, author of the classic *Visuddhimagga,* the Path of Purification, states that with access concentration the hindrances eventually become suppressed. The defilements subside, and the mind becomes concentrated. What is referred to here is the temporary subsiding of the five hindrances: of (1) sensual desire, (2) ill-will, (3) drowsiness and lethargy, (4) restlessness and worry, and (5) doubt, or fear of commitment. The five hindrances are recognized in all of Buddhism. These are afflictions that are bothersome in daily life, and certainly in meditation. When the mind is dominated by them it is not fit for concentration, and the cultivation of insight. According to Wallace, the actual accomplishment of this ninth stage, attentional balance, involves a radical transition in one's body and mind. One is like a butterfly emerging from its cocoon, one experiences mental pliancy, and physical and mental bliss. The mind gets fit and supple like never before. An interesting way of describing it I find with Gyatrul Rinpoche: "Those are indications by which you can know that quiescence has arisen in your own mind-stream. When you have achieved that, you will never need to tell anybody 'give me space,' and you will never feel lonely." And he adds: "Wherever you place your attention, it will be imbued with the quality of serenity, like a candle flame that is not flickering due to wind."[9] As to trust and dust: this sounds like quite some dust has left from the eyes. Certainly there is a clearer view!

A person may proceed beyond access concentration to the actual state of the first meditative absorption or stabilization (as we saw represented on the Elephant Path, by the elephant walking on the multi-colored bands). This comprises the suspension of the five hindrances and the manifestation of five factors of stabilization that have gradually been building up. The first is mental application, referring to the process of applying the mind to the object. The second is examination, leading to the mind remaining on the object. The third factor is joy, this may begin to express itself in the body as warm tingles, culminating in a sense of bliss that pervades the entire body. The fourth factor is happiness, in the sense of a deep contentment, more tranquil than joy; it arises as the mind becomes harmonized and agitation has gone. The fifth is one-pointedness of mind: here the mind is wholly and contentedly unified on the object.[10] All these practical lists in Buddhism! Next to this proceeding to meditative absorption, access concentration can be a good foundation for the practice of vipashyana insight meditation. Moreover, access concentration may be needed to progress on the path of the *bodhisattva*, the path to perfect Buddhahood. It is the *bodhisattva*'s task to compassionately help beings, while maturing her or his own wisdom. Certainly, a very high degree of mental health, fitness and pliancy is welcome for developing a sufficient basis for contemplative insight and for the cultivation of great love and compassion.

Absolute and relative

When pondering these expressions "attaining substrate consciousness," and "achieving shamatha," the first raised many questions for me. How to better understand the meaning of substrate consciousness? As briefly referred to in Chapter 2.5: substrate consciousness is seen as a subtle continuum of mental consciousness, the relative nature of the mind. The Dzogchen tradition draws a distinction between the substrate consciousness (*alayavijnana*) and the substrate (*alaya*), which is described as the objective, empty space of the mind, as a "vacuum state." This vacuum state, according to Wallace, is immaterial like space, and described as a blank, unthinking void into which all objective appearances of the physical senses and mental perception dissolve when we fall asleep. It is out of this vacuum that appearances re-emerge when we wake up. For me it has been clarifying to see and get a felt sense of this distinction by way of viewing the not-oneness of substrate and substrate consciousness. This connects with the way we can understand what is referred to as relative and absolute.

Regarding *absolute*: the Great Perfection is said to be the unity of absolute space and primordial consciousness. In this, absolute space refers to the fundamental nature of the experienced world, and primordial consciousness is the fundamental nature of the mind. The Great Perfection entails the *nondual* realization of the intrinsic unity of these, transcending all distinctions of subject and object, mind and matter

Regarding *relative*: the substrate consciousness is different from the substrate that it ascertains, so here is no unity, here is *duality*. While the substrate consciousness is qualified by distinct experiences of bliss, luminosity, and nonconceptuality, it is ascertained when the mind is withdrawn from the external world of the senses, and it is bound by time and causality, specific to a particular individual.

The unity of absolute space and primordial consciousness (absolute) is also imbued with the qualities of bliss, luminosity and nonconceptuality; yet, here they are not present as distinct attributes (as they are in relative, substrate consciousness). While in the relative sense, with attaining substrate consciousness, the experience of bliss, luminosity and nonconceptuality is limited and transient, in the absolute sense of primordial awareness, these are inconceivable, boundless, and timeless in scope. Realizing the substrate consciousness by achieving shamatha leads to a temporary suppression of mental afflictions, in other words: the hindrances go dormant. However: when realizing primordial consciousness, it is said that all hindrances, mental afflictions and obscurations are eliminated forever. [11]

Interestingly, and happily, we don't have to wait to taste aspects of experiences in the sense of decreasing hindrances and increasing felt sense bliss, luminosity and nonconceptuality, until we fully attain shamatha. While doing the shamatha meditations for a somewhat longer period of time, I could experience a lessening of the hindrances. In the beginning there was a lot of drowsiness, and excitation; they gradually diminished, as did some wishes and desires, there was more contentment. Also the gradual manifestation of the five factors of stabilization could be felt: next to concentration getting better, I've been so aware of a sense of deep joy that was there, all the time. This joy came to underlie passing emotional upheavals, as described in my diaries ... on the way to increasingly fully settling of the mind in its natural state, in the sense of "resting" in substrate consciousness as the vantage point.

4.4 The vital role of shamatha

While it has often been a painful confrontation for us as participants, it became so clear how thoroughly habitual and immense this unrest and agitation in our body-mind was, how much work still had to be done. Alan suggested that three months is just an eye-blink in comparison with what is said on the duration and intensity needed to fully achieve shamatha. This, he said, might take many more months or even years. As there is a widespread consensus concerning the vital role of shamatha in Buddhist contemplative practice, one would expect that it would be practiced in a quite disciplined way and that many contemplatives would fully achieve shamatha. Oddly enough, Wallace states, a strong tendency seems to exist among Tibetan Buddhist contemplatives to marginalize shamatha in favor of more advanced practices. Tsongkhapa commented on this oversight in the fifteenth century when he said, "There seem to be very few who achieve even shamatha," and Dzogchen master Düdjom Lingpa commented four centuries later, that only very few appear to achieve more than fleeting stability, among unrefined people in this degenerate era. Interestingly, many centuries earlier, it was already advised by Buddhaghosa to the monk-meditator that " ... he should avoid a monastery unfavorable to the development of concentration, and go to live in one that is favorable." Which apparently was not self-evident. [12]

Both the Theravada and the Indo-Tibetan traditions of Buddhism agree that the cultivation of shamatha leads to an experiential realization of the ground state of the mind or psyche. While being referred to in later Buddhism as *alayavijnana,* as we saw, this ground is referred to in Early Buddhist literature as the *bhavanga*. This is seen to literally be "the ground of becoming," and it supports all kinds of mental activities and sensory perceptions. Continuing on musings, as described earlier in relation to the Questions and Answer sessions: I wonder about the nature of this ground, is this like a "stream of consciousness," is it on-going, and continuous during experience, during my life? Will it stop when I die?

Theravada commentators insist that the *bhavanga* is an intermittent phase of consciousness, which is interrupted whenever sensory consciousness or other kinds of cognitive activity arise. Peter Harvey, Buddhist scholar, clarifies: *bhavanga* is seen as the resting state of consciousness, and this level of mental functioning is seen as being constantly flicked in and out of during waking consciousness. Parallels are sometimes made with a TL-lamp: the pulses go so quickly that there may seem to be continuity; still, there are pulses. On the other hand, Sopa refers to *alaya* consciousness in the context of later Mahayana schools that had their theory on the *continuity* of an individual consciousness. In referring to the view of contemplatives practicing in the Dzogchen Mahayana view, Wallace states that the substrate consciousness consists of a stream of arising and passing moments of consciousness; in that sense it is not permanent. Also, it is conditioned by various influences, and in that sense it is not independent. However, these contemplatives do regard it as a continuous stream of consciousness. It is from this stream that all mundane cognitive processes arise. [13]

There's a vital role for substrate consciousness in the understanding of the range of mind and consciousness states and awakening, and a vital role for the practice of shamatha as a complement to insight practices. But there may be misunderstandings. Coming back to relative and absolute: while contemplatives who have realized the substrate consciousness through the practice of shamatha describe how it is imbued with the attributes bliss, luminosity, and nonconceptuality, it is a grave mistake to take these temporary experiences for genuine realization. In that case practitioners make the mistake of taking the substrate consciousness (the relative ground of the psyche) for the absolute, ultimate nature. Dwelling in this relative ground does not liberate the mind of its afflictive tendencies or of the suffering that comes from them. As Thrangu Rinpoche, contemporary Tibetan Buddhist teacher, states: getting to know the nature of consciousness in its relative ground state, as an individual stream of consciousness that carries on from one lifetime to the next, does not illuminate the nature of reality as a whole. There can't be enlightenment in the relative ground state; but the stream goes on to the next lifetime. [14] So, these are some Buddhist responses to my questions: according to many, there is a (seeming) continuity in the relative mind stream, and it is not only seen as being there through life, but also through death …

4.5 Mindfulness and introspection

The notions mindfulness and introspection – also referred to as awareness – having been mentioned in passing in Chapter 2, deserve more attention in this context.

Buddhaghosa, the great systematizer in Early Buddhism, drew the following distinction between mindfulness and awareness, in relation to concentration. Mindfulness has the characteristic of remembering, its function is not to forget. It is manifested as guarding. Full awareness has the characteristic of non-confusion, its function is to investigate, and it is manifested as scrutiny. Jumping from fifth to fourteenth century: Tibetan master Tsongkhapa speaks of two requirements for shamatha practice: (1) a means so as not to be distracted from the meditative object, and (2) a knowledge of the true state of affairs: distraction or non-distraction, the threat of distraction or the non-threat of distraction. The former is mindfulness and the latter is awareness. [15] So, both Buddhaghosa and Tsongkhapa refer to mindfulness (Pali *sati,* Sanskrit *smrti,* Tibetan *dran pa*) in terms of remembering and non-distraction, and to (full) awareness (Pali *sampajana,* Sanskrit *samprajanya,* Tibetan *shes bzhin*) as standing for non-confusion, knowledge of the true state of affairs, also referred to as clear comprehension or intelligence. As these notions, mindfulness and (full) awareness are so central in our shamatha discourse, they are presented here in three language tastes.

We saw how Alan Wallace defined them, in a practice way, with us in the project. He described mindfulness as the attending continuously to a familiar object, without forgetfulness or distraction. This non-forgetfulness includes retrospective memory of things in the past, prospectively remembering to do something in the future, and present-centered recollection in the sense of maintaining unwavering attention to a present reality. For another contemporary voice: Geshe Gedün Lodrö, Tibetan Buddhist teacher, states in the same vein that mindfulness has the feature of an object of observation with which one is familiar. Its subjective aspect, or way of apprehension, is the non-forgetting of the object of observation.

Introspection was described by Wallace as the meta-awareness that allows for the quality control of attention. Connecting with this, Lodrö states that at the time of working on stabilizing meditation, "with a corner of the mind" one remains cautious about becoming lax. Very refined distinctions are made between different degrees of laxity and excitation, with different terminology, and in great detail. Translations refer to distinctions made between, on the one hand, fading, drowsiness, dullness and laxity. On the other hand, there is scattering, restlessness, excitation and excitement. [16]

The question comes up: regarding mindfulness, is this the same kind that we hear of in contemporary applications like Mindfulness Based Stress Reduction (MBSR) and Mindfulness Based Cognitive Therapy (MBCT), or in vipassana centers? I had been involved in working in this mindfulness-context, and also I had been practicing at a vipassana-center. Now, being with Alan Wallace, I heard a way of defining mindfulness that differed from the ones I was familiar with. Which made me do some investigation.

Musings on mindfulness

Contemporary mindfulness based approaches profess to base their understanding of mindfulness in early Buddhist vipassana insight meditation. Do we find this back in their descriptions? Jon Kabat-Zinn in 2005 describes mindfulness as "moment-to-moment, non-judgmental awareness, cultivated by paying attention in a specific way, that is, in the present moment, and as non-reactively, as non-judgmentally, and openheartedly as possible." It seems that in this definition there is less emphasis on the direction of attention, and more on the quality and awareness. In contemporary mindfulness approaches, based in vipassana, the term mindfulness seems to be used mainly to refer, not to the focusing aspect of mind, but rather to the introspection, the meta-awareness that surveys that focus and its relation to the intended object. Which is in contrast with the Tibetan use of the terms of mindfulness and awareness, where, as we saw, mindfulness refers to attention or focus, while awareness refers to a faculty of mind that surveys the mental state at a meta-level.[17]

For me, after initial confusion, it has been clarifying to understand these various views. It is an interesting difference in felt sense, between feeling confused, and seeing what the confusion is about, from a meta-perspective. The feeling of confusion, I realize, was also connected with a sense of identification with the meaning of mindfulness that was mostly on the foreground for me, at a given time. When we developed and gave mindfulness courses at the center where I worked with traumatized refugees, of course there was the mindfulness meaning as given by Jon Kabat-Zinn, with whom I had taken a course. Later, there was the meaning of mindfulness, as presented in a vipassana group where I took part. However, when participating in the Shamatha Project, there was the Tibetan mindfulness context. And then it's so good to know: it's not about terminology ... It's about meaning, felt sense, about embodying, living it. There's a joyful sense of relief in the integrating practice of developing a wider meta-perspective ...

Diary, 2008

I feel so happy to be more intimate with, to be better able to dwell for shorter or longer periods in this meta-position with equanimity and clarity ... while there sometimes is such an overload of mental chit-chat, this and that, pondering, comparing etcetera going around in the mind. Being home in this dynamic, knowing, radiant purity of this meta-position. This "meta" in felt sense, connecting with *metta,* Loving Kindness.

Mindfulness, bare attention and Four Applications

My investigation then turns to the notions: mindfulness and what is named "bare attention." Generally, a distinction is made between "mental engagement," or "bare attention," on the one hand, and mindfulness, on the other hand, but sometimes in mindfulness circles I heard the two terms used as if referring to the same thing. Bare attention here refers to the initial split seconds of bare cognizing of an object, before one begins to recognize, identify, conceptualize. This connects to the Question and Answer sessions, described in Chapter 2.5, where the importance of this brief moment was emphasized, and where we learned that shamatha training generates the vividness, necessary for discernment of this brief moment. Bare attention is seen as an "ethically variable universal mental factor," in terms of early Abhidhamma Buddhist psychology. A mental factor is a state associated with consciousness, an aspect of the mind that apprehends a particular quality of an object. A distinction is made between ethically variable, unwholesome, and beautiful mental factors. This bare attention, as a universal mental factor, is common to all consciousness. Mindfulness, as a beautiful mental factor, in that sense connects just with a beautiful quality. So, when there is bare attention, this needn't mean that there is mindfulness. By the way, relating to trust: mindfulness (Pali *sati,* Sanskrit *smrti*) is listed as the second of the beautiful mental factors, and faith or trust (Pali *saddha,* Sanskrit *sraddha*) is named as the first! [18]

And how does mindfulness relate to concentration-calm, on the one hand, and insight practice on the other? While in his original description in his chapter involving concentration Buddhaghosa connects mindfulness with remembering and not forgetting, in a chapter regarding understanding and insight, he expands the meaning. At this place he describes mindfulness as follows: " ... not to forget ... manifested as guarding, *or* it is manifested as the state of confronting an objective field. Its proximate cause is strong perception, *or* its proximate cause is the foundations of mindfulness concerned with the body, and so on ..." Here the first part of the sentence refers to concentration practice, the second part, after "*or*," to the "foundations" of mindfulness and insight practice. This regards the basic framework for developing insight practice: the Four Applications of Mindfulness, concerning the body, feelings, states of mind, and reality patterns. With the Four Applications insight practice, rather than focusing on one chosen object like in concentration practice, the attention is opened out so that mindfulness calmly observes each passing sensory or mental object. [19] So, instead of focusing (in concentration), there is now an opening up (in insight practice). This, too, has been clarifying for me. Here we see some subtle divergence taking place for the meaning of mindfulness, toward either concentration-calm or insight context.

Chapter 5

Three months of sitting with Settling

> *Often coming up, as objects: feelings and insights about the practice, the intelligence in it, in refining attention. So great refining still to go! The practice: not taking anything away, or adding anything from outside.* (Diary, 5.4, nr 38)

After the historical and theoretical explorations in Chapter 4, let's return to practical shamatha.

In what follows, some impressions are given from diaries that I kept during the three months of the Shamatha Project expedition in 2007, and from audio and video data. Reading these now, I feel touched by their innocence, in a sense that, at the time that the diaries were written, I didn't have much knowledge about states and stages, about what could (or "should"!) show up. I was rather fresh to the practice, and not so familiar with the texts and theories laid down in this book, at the time that most of these diary notes were jotted down.

The data below comprise just one example of how one participant – me – was present in the Shamatha Project process. These data are presented here to, hopefully, convey some of the inner feel, inner taste, and of the dynamics of a possible evolving process. Even while the data are first person, a third person approach will resonate sometimes in the descriptions. Next to the general flow of experiences, I will also describe turbulence and turmoil. These were the *nyam*, transient anomalous experiences in the process, that especially came up with Settling the mind in its natural state practice.

While up till here all guided meditations have been taken from tape recordings made of Alan Wallace's instructions during the beginning phases in the Shamatha Project, from here on some more variety comes in. Some, again, are Alan's instructions from later in the project, some are presented in a somewhat more loose way, breathing his style of guidance, and inspired by some elements in his writings. As Alan invited me, in a later shamatha retreat, to present guided meditations, I also include two of these.

5.1 Evolving practice, impressions from empirical data

In the sections that follow, I have grouped the data by month. Roughly, the first month is September 2007, from the start of my participation, at the beginning of that month. The second month is October, the third month is November to the end of the retreat, in the beginning of December. Mostly, the sections are in a sequential order. Selected diary fragments from personal diaries are sometimes complemented with information from the Daily Experience Questionnaires. The personal diary notes show examples that are almost real time jottings down, with as little processing and conceptual overlay as possible. I present some slightly (milliseconds) later "second-order associations," ("next"), in italics. Sometimes a comment or explanation to the reader is added, just for this writing.

In this selective diary presentation of experiences, for the beginning phases I tentatively give special attention to material relating to the object of meditation, to the subject meditating, and to the process of meditating, including the quality control of monitoring and awareness. In other words: attention is given to the observed, the observer and the process of observing. This way of structuring the material, for the start, offered itself in a natural way. The audio recordings (once in two weeks) present some more working through. The video interviews, once a month, include some interpersonal, intersubjective dynamics. Only a few audio and video quotes are presented, and in paraphrasing, as it turns out no personal data that can be traced to a specific person, are allowed to be published out of the project-database.

The materials are generally presented starting with the most direct experience, gradually moving toward materials that show some more digestion of the experiences. In a way of direct experience, there would be logic in placing Chapter 5 before 4, as, when I went through the process, the knowledge of Chapter 4 was not with me in an organized way. Still, for ease of reading the present way seems more appropriate.

At the end of the description of experiences in each month, I present a section on turbulence and turmoil. Dream reports, giving a specific window on inner processes, are included. I give a number to each information item, for easy reference.

5.2 First month meditation practice: object, subject, and monitoring

Here are some examples regarding the object of mindfulness, in the first month:

> The object of mindfulness: the space of the mind and what arises in it …
>
> I'm so aware that, with a little stress (it happens, even in this environment!), forgetfulness increases, as do distraction and grasping. It's the way the mind-brain is wired ... Also, I'm so aware of this tendency, next to this un-mindfulness, to want to "think out a thought," with some grasping. Feels like sort of a micro-anxiety of forgetting; a tenacious compulsive habit, not trusting that the thought will come up again, or not – it's all conditioned arising. The grasping is just habit. At times there seems to be a sense that this thought may be so important, that it "has to be completed." (1)

> Near to my room is a parking place with cars arriving, leaving. While during the first week this was just disturbing, distracting, I can now use this in a reframed way as "mindfulness bells," reminding me of being mindful, grist for the mill! Right away I feel some relaxation.
>
> Of course, there's manipulation in this, placing something I

> don't like into a different context, it is not just observing, but it is "training wheels," as Alan calls that. He quotes Lerab Lingpa in saying "be your own mentor." Let me just explore this for a bit. It's also an exercise in shifting perspective. (2)

Regarding the subject, in the first month:

> Yesterday I did peaceful ongoing hours of practice, experiencing stability and vividness increasing. Today a day of soberness, sadness. Of course there will be ups and downs. But this also feels like a shift, in these two weeks, something changing in me, to more spaciousness … with joy in soberness and simplicity, and some as-it-is sadness about the many years of subtle agitation, not really questioned, taken for granted. (3)
>
> I'm aware that the range of contraction and relaxation is so wide. Perception is developing into greater detail. There's so much with *energy* in this practice! Noticing that any distraction, grasping, any thought influences the breathing. Meditating now for a few weeks: this habitual state feels so different from any "natural state," whatever that may be. Habitual state feels as imbalanced, in the sense of a slight (or not so slight) habitual tension. I "feel" with all inner senses: see, hear, smell, taste – how this contraction closes off, shuts down. Relax, release ... (4)
>
> *Next: perception-emotion-association: feels like an imbalance of the autonomic nervous system, in the sense of habitual tension (the sympathetic aspect), less release (parasympathetic). Contractions narrowing, closing off energy channels, free flow ...*

Regarding awareness and monitoring, first month:

> Today much distraction and restlessness, in a vicious circle with sort of an ambition for the practice. It helps me to try to release the energy behind what feels like an urge for and at the same

time resistance against the practice, and not bother about the restlessness per se. (5)

I'm aware: while shamatha is mainly observing, in a way "receiving," there is so much micro-"doing." I can hardly receive this gift yet, in just being, in this selective way. Receiving, so difficult for some part in me that doesn't have a clue of non-doing, or being. This part in ignorance, wondering if this is "waiting" – for what? (6)

As to monitoring: I feel like I'm orienting myself in spaces. Space of body-breath, space of mind, space of awareness. Many metaphors come to mind. As if, initially, what I'm focusing on is like photography "layers" of depth in focus. The lens of a camera, focusing, and tuning into a slice of reality that becomes sharply represented in experience, ready for being observed. It's a limited metaphor. Yes, with shamatha there is selecting while tuning in with a certain band width that, with increasing precision, will increase – later up to all-inclusiveness. The space of awareness, always everywhere ... (7)

Here are some reflections to these first person experiences.

In the first month the object of mindfulness is in the foreground, gradually coming in with more detail and refinement. The monitoring and also the style of monitoring are objects to be observed, by themselves. I'm aware that habitual subject-object relationships present themselves, and that conditioned patterns are getting more visible. Which is not just so pleasant.

I've strongly felt energetic embodiment dynamics. Although while practicing Settling the mind there is "no interest" in physical experiences, there are kinesthetic and mind-body propriocepsis-energetical sensations. These, with inner mind-body perception, cannot just be named physical. Looking back: also I noted synesthesia, perception of stimuli in various sense modalities at the same time. I'm seeing sounds, I'm hearing colors. And there is a sense of inner seeing, that's more a form of knowing.

The subject per se is not given much attention, I'm not much observing who is the one who does this. While my person, my mind, my "self" is not questioned in the beginning phases, and mostly taken for granted, I'm aware that my inner formulating shifts somewhat from "I" to "there is ..." with a functioning-orientation. Like: not I select, but: there is selecting. This has a different felt sense. The observing is mostly done by me as the observer, the watcher, and there is much discursive thinking. Instructions are followed, with increasingly strong mindfulness, and with a relatively coarse sort of monitoring of the process.

When looking back, and writing this, I'm so aware of the fact that the wish for maximal directness and "real-time-ness" meets clear limitations. Jotting down includes a time lag, there is translation of wordless feeling tone, thought-flash, emotion-wave, into these clunky stammering words. Fingers pointing to the moon.

Meditation turbulence, first month: nyam, lessons in diving

In Chapter 2.5 the phenomenon *nyam* has been briefly addressed: a transient anomalous psychological or somatic experience, catalyzed by the practice of meditation. In that chapter I presented some of Alan's advice as to how to handle strong emotions, anxieties, and confusions. The baseline was: with physical phenomena that may represent medical problems, we don't take risks: a nurse and a doctor can be consulted. Otherwise, when there is *nyam*, just "watch your mind heal!"

Here are some examples of meditation *nyam* and turbulence during the first month:

> I'm seeing how I often place the idea that something is solid over something that I may feel, in the tactile sense, as if it comes from an old databank. When I look: "this is my index finger," then

there can be an imaginary seeing it as a solid body part. There's quite a difference between a conceptual imagination and just the felt sense, "being it." Sometimes I know my index finger as empty of substantiality. There are molecules, or density in the energy-field, and there's no "ownership" of this density. The sense is very natural. The index finger is all-over dependently originating. Today, after the session there was a feeling of derealization, a sense that this finger, and objects in the room are unreal. There was sort of distancing, as in the beginning it is so "other" than I conventionally experience this.

There's more and more loosening of unconscious compulsive projections, seeing my entanglements, transferences. Seeing how I tend to displace feelings, patterns, mostly from the past, now to persons and things in the present. There's less solidity and substantiality in that as well. (8)

Comment to my using this kind of terminology: it feels like a mixed blessing, to be in the psychology-psychiatry profession! Still for me, these words are familiar, so they come up quite easily, without much processing or conceptualizing. They don't carry much burden, they are micro-descriptive of familiar dynamics. Explanation: transference can be seen as the redirection of feelings and desires and especially of those unconsciously retained from childhood, toward a person in one's present life.

The most clear *nyam* that I experience many times is that my hands feel like claws, grasping, with clamping bent fingers, and as if having a greater size. An awkward and uneasy feeling. Coming out of meditation I'm often surprised, seeing their familiar form. Sort of perceptual distortions in the tactile field. (9)

Other *nyam* I recently experienced:

Last few days: visual *nyam*, seeing a dark form moving on the floor, thinking it's an ant, but no. Speckles of light, vibrancy during day and night

> Sudden hot flushes
>
> Muscular shocks and pains in my legs
>
> Last few days: finding myself suddenly speaking in Dutch (native language) to others. The few words that we speak here! As if coming from a deeper, older layer. (10)

In what follows I present some typical dreams that I had during this first month. In the Daily Experience Questionnaire section about dreams that we fill in early morning, a number of requests are mentioned. The first is to, when you remember any dreams from last night, describe each briefly and say briefly how each made and makes you feel (anxious, sad, puzzled, joyful, etcetera). Discuss up to three dreams.

The following requests are then added, for every dream: to describe any *recurrent* dream signs: this is about emotions, people, situations, places, or activities that occurred more than once in the dream (or across dreams). Additionally the request is to report any *anomalous* dream signs: events in the dream that could conceivably occur in the waking state but would be very odd if they did. Finally, we are requested to describe any *impossible* dream signs: referring to events in the dream that couldn't occur (as far as you know) during the waking state.

I had many dreams in relation with the Shamatha Project. Two are presented here:

> We get sort of lessons in "diving" in shamatha-substrate consciousness. There is an instructor with a long pole and hook, and we pass by him – like children, as a group in swimming lessons. His pole and hook are there to save us, and if necessary, drag us up. We learn skills, and are all the time reassured by this instructor in white clothes, with going deeper. It feels kind of exciting, also normal, necessary, like swimming lessons when I was a child. (11)

> I'm participating in a thorough meditation retreat. I feel it is important to have as few burdens, obligations, stuff as possible, to be as "bare," naked, sober, simple as possible. I take one piece of clothing less all the time, everyday, one day no shirt, next day no scarf as well. I'm aware that my head hair has been shaved (which is the reality).
>
> Mentioned in Daily Experience Questionnaire:
>
> As recurrent dream sign: often, in my dreams there is the theme of "radically simplifying my life" – here is one variation of it.
>
> As anomalous dream sign: taking one piece of clothing off, all the time. This could also be one shoe, trousers. In the dream this did not feel pleasant, but evidently it just had to be done, as I made this commitment. (12)

Looking back on diary fragments of turbulence in the first month I note down some impressions:

The dynamics are as described in general about this month, but these turbulence-sections seem to have some sharper edge. The intensity of entanglements is felt and described as stronger. I name exceptional somatic experiences, perceptual distortions in somatic and psychological sense, and de-realization. I have had many dreams, and quite a few of them are about the project.

5.3 Second month meditation practice

Up to the second month. Here are some examples, regarding the object of mindfulness:

> In comes a feeling of falling, with some subtle sense of gravity ... uncertainty. Like descending through layers, earth layers like in archaeology. Now: through awareness. While observing objects of mindfulness, not only: what's arising, also: what seems to be arising while there is "descending." (13)
>
> *Next: Freud used this metaphor of archaeological layers.*

Not losing the thread of continuity. The work is in the refining in completeness: thought bubbles in the periphery, in the background. Observing and "sensing" these, more and more quickly. Taking greater, closer, interest in the object, with greater intensity. Balancing with relaxation, loosening up, balancing on the edge.

What's the difference of this observing, on the one hand this "seeing" emotions, thoughts, and on the other, insights in insight practice? This is not vipassana insight practice, with exploring the "Three Characteristics," the reality of impermanence, suffering, non-self; or emptiness of self. I don't explore, inquire. Still: it's so rich, this psyche, these mind processes, that just manifest these characteristics! Thought arising and dissolving, impermanent. A pleasant feeling that doesn't stay, impermanent. Grasping making things worse. The space of mind: no center, no periphery. Insights about mind-aspects that usually are subconscious. (14)

Regarding the objects that arise: now, after six weeks of retreat there's so much more silence in the space. There may just be vibrancy. At other moments I can perceive arising objects separately, and also I see connections in chains of arisings, in Settling the mind practice, one inviting the other. For instance, one example, viewed with a magnifying glass as to space and time:

1. Arising: remembering, this morning Jill having received chocolate bars by mail. This comes up as a mental image, with mind-body bringing in the physical. Desire, inner taste ... Flash of denial, I don't want to feel this longing. Non-grasping, back to space of mind.

2. "Burning" envy! Pang in the belly, I want it! A flash of: I'll pick it away, I don't want her to have it, don't like her. Again, mind-body, felt-sense.

3. Inner response: Oh, I don't like feeling envy; a critical voice,

mental overlay, contraction.

4. ... But it's there, all inclusive. Acceptance, a little less limitation-contraction.

5. Such it is, these are all observed, to their nature. It's just passing feelings. Witnessing, no judgment. Then more freedom-relaxation. In this witnessing: more true acceptance, soothing. (15)

Settling practice gets more and more juicy. There is more immediate lived experience, more adventure. There's stability in not losing the thread of continuous mindfulness, for longer stretches. At first, continuity with object of mindfulness in the background is reached, then also more in the foreground, more complete, with less noise. I feel so happy "I got here!" Objects of mindfulness are perceived as connected with all the (inner) senses, sometimes specifically tactile-sticky, smell-stinky, the foam that keeps things clunky, untransparent, vague, like pond scum, a surface layer. There is a feeling of floating in it, sometimes it feels like with the same specific gravity – which may stand for: identifying with thoughts. Then, no floating, no subject-object, just being. (16)

Here are some fragments regarding the subject, in the second month:

Tuning in with breath meditation, I'm more surrendered to "being breathed." Sometimes: witnessing every breath, every unique dynamic in it. The freedom, non-attachment in the witnessing of some impression is there also outside meditation, the surprises, with unexpected sounds, views. I'm feeling new, fresh. (17)

More often in meditation questions come up: like who or what is doing the concentration? What is mind? Alan invited us to ponder questions like these for a while, just in the beginning:

does the mind have a color, a location? Are there boundaries to the space of the mind? A center? Can the mind be damaged? Can we be sure that our mental space is private? Is there a "membrane"? What is it that knows, or does not know? (18)

Next: During meditation Alan just quoted Padmasambhava: alternate between observing who is concentrating inwardly and who is releasing. He said something like: this is a kind of questions that needn't be verbalized, and certainly not answered in that sense. In this context the questions are brought in to counteract laxity. We are postponing questions like: who am I, to vipashyana.

More trust, feeling more safe internally. Feels like a "breaking the barriers" of anxiety, inwardly. Little steps, step by step, deconditioning by deconditioning, mind moment by mind moment, molecule by molecule. Feeling more at home, familiarizing. It seems to coincide with this wondering about the me-feeling. Me-feeling more often is not there. Space of the mind: no center, no periphery. No breath, no breather, no breathing, just "being breathing." (19)

Next: it makes me think of "priming," in the sense of attachment style: shifting to a more "acquired secure attachment style," with more trust. Reaching back to wholesome relations, deep brain structures, pre and non-verbal. Maybe also: some sense of reconnecting to substrate consciousness, where pristine awareness shines through. Explanation: Priming, in research context, refers to a situation in which one is induced, consciously or subliminally, to access mental circuits that are associated with security. Attachment style is (primarily) formed early in life, as a result of how a child is responded to by, and how it interacts with her or his primary caregivers. It can be seen as a style in relational being with oneself and others. Some plasticity in style is possible, research shows!

"The less dust, the more trust" as a mantra, this comes up on a walk. A great expression about people who have "little dust in their eyes." Less tendency toward micro-control, anticipation and checking. I'm so aware of these micro operations of habitually locating myself in checking coordinates where-what-who am I, contextual, in dependency. So often not being "now": as if always having to check forward, backward. There must be an underlying anxiety, so habitual, so craft-fully covered. Sometimes in practicing Settling the mind – space of mind, and in Awareness of awareness: some flashes of anxiety for this empty space of mind, awareness. (20)

I'm so aware of the elegance, and self-reinforcing ways in the method. Occasionally there may be some grasping and excitation in my joy and enthusiasm. Sometimes it feels like almost "too much." Sort of inner thrill, breathlessness. Dazzling, this large space. It may give some anxiety. Today I chose to spend quite a few hours safely in the supine position in bed, no external demands. Body-mind feeling: small, fragile.

For the first time here: this clear feeling of loss of control ... Can I stay in the curiosity, non-doing, not-knowing? This is what I came for! Surrendering, observing, while grateful for the safe environment. Later, in the afternoon: more quiet, more balance. (21)

The non-grasping! Thoughts feel so heavy, verbal thoughts and words: heavy and slow, and keeping one (me) to the surface, covering off the subtler lightness glow and freedom flow.

There is a "descending" that feels at first as if helped by gravity, by a body. Then: it is descending in more lightness, thinner energy, more light. And then there is no gravity or direction anymore, up and down is preceded, transcended. And there's this underlying experience of gentle joy, all the time. (22)

Thought-grasping leading to contraction, subtly palpable in my hands. Some thinking-excitation is there, irrespective of the content of the thinking felt as pleasant or unpleasant. This thinking-excitation seems to create this "me," the subject.

I realize that grasping acts "to its best knowledge" and intentions, abilities, just with its degree of ignorance – like all beings, all phenomena do! This grasping wants to make "me" happy ... but there is no "me" in the way grasping grasps – and that's why it desperately conceives of its deluded task-description as grasping. Only grasping doesn't know that! (23)

... And these are some diary remarks about monitoring, in the second month:

The detoxification process ... It feels like a micro process of *"yo-yoing":* there is some descending – to stammer in this language of locality, gravity. There's a loosening, and when this becomes too anxiety provoking, then a bit "up." And descending again, and refining, with micro survival reflexes, subtle inner bio-feedback in doing the practice. A sense of melting, overall. Little spikes of moments of anxiety: then some arising-being-further dissolving, arising-being-dissolving ... first with larger amplitudes, then amplitudes getting smaller, and more dissolving. There is more and more relaxing, release.

These are the unruly guests, the guests that have been welcomed, anyhow, by the gracious host, again and again. They don't need to demand attention anymore. (24)

Next: feels like a process of subtle systematic desensitization. Explanation: systematic desensitization, in Western psychological context is an approach based on the behavioral principle of counter-conditioning, whereby a person overcomes non-adaptive anxiety that is aroused by a situation or an object by gradually approaching the feared situation. The person is in a psycho-physiological state that inhibits anxiety.

Next: "the gracious host amidst unruly guests." Explanation: Sogyal Rinpoche refers to awareness as the gracious host in the midst of unruly guests. Wallace often quoted him, in the sense that guests come and go, they may have food fights at times, but the host is calm and serene.

So many new insights, a flow of little "ah"s!

In my amateur-physiological imagery and also felt-sense: like little "arcs" in dynamic neuronal networks, connecting, flashes ... Also: deepening relaxation, how about beta to alpha, to theta, to delta rhythm in EEG? How about gamma? Amazing, this feeling of getting out of a habitual agitated "groove" when in the Settling the mind practice, and coming into a field, a channel of quiet liveliness. Wondering: how might this be coinciding with different "band width" on EEG?

There's little mental elaboration, and far less conceptualizing. Also, sometimes there is a feeling of: transcending (with including) the physical senses – they are not gone, they are in the background – not identified with. They are absent as long as there is no attention going to them – a glimpse of "body falling away" as named in Zen. In the mind-range: no "mind" in the old sense, yet heightened presence, in this holistic energy field. Body and mind falling away ...

Intuition, more clarity and presence. On my walks in the mountains, all forms and contours seem to be more clear, fresh, rich of contrast and not so demarcated at the same time. (25)

A dazzling week – so I tell Alan in the weekly interview, near the end of the second month. Insights, exuberance, and deep quiet. Processes of anxiety, feeling of loss of control ... Insecurity, wishing his approval? And extreme tiredness.

I come with the urgent question: how to handle these insights, without grasping, and disturbing the meditation flow? Short sessions and then jot down some notes? Alan comes with some advice: 1. Having trust; it can't get lost, it's (in) you! It will come up in its own time. 2. On the other hand: it may be good to do two

> sessions a day of insight practice, vipassana, deliberately. Two times thirty minutes, notepaper and pen allowed. So, then there is some space for shorthand words, and later you can elaborate. I tell him that, actually, already I found a way of "writing" a few words or signs with my finger on my thighs, a sort of anchoring. It helps, retrieval seems more easy when I have stored this information with multiple senses, including body-tactile movement memory. (26)

In an *audio* recording during the second month, I name a special aspect of practice opportunity: the sense for me of having come here while not knowing any one of the participants. It's the external aspect of the opportunity of being completely new and fresh, as far as that can be. It gives some sort of curiosity also of observing and witnessing who is coming out as "me," as this body-mind. (27)

In the *video* interview I describe ways of experiencing space. I refer to being aware *how* much energy is going into sort of keeping this "self" in place. This endless splitting, vortexing thing to oversee, be prepared ... for what? Feeling anxious micro survival reflexes, "not leaving myself alone." (28)

Looking back at these impressions in the second month: in this phase, there is increasingly less grasping to objects. There is hardly a discontinuity, and more completeness in being on the object. I'm aware that subjectivity comes much more to the forefront, in a way also as an object of mindfulness: who, what is meditating? Questions that come up, not to be answered, but supporting the stability and coherence in the practice. With that, there is now much stronger mindfulness with seeing objects arising in patterns, and seeing the involuntary inner reactivity, in patterns as well. At the same time: subject-object demarcations are fading, less duality. Underlying is a gentle joy.

While writing now for a book, looking from a meta-meta position, again I feel this joy – it has deepened, it is reverberating, multiplying within these exponential meta-perspectives.

Meditation turbulence, second month: "I do things wrong so I exist"

The second month really taught me about nyam! Some examples of meditation turbulence experience in this month are the following:

> I get to understand: as soon as looking from conventional mind, there is non-normal, anomalous experience catalyzed by meditation, this is called *nyam*. As in Settling the mind practice the "mind is reverting to the substrate," and takes the perspective going with the level, this can give insights in reality that may be "anomalous" for conventional mind indeed. When we are "on the way," gradually our view will change. Seen from a more subtle and healthy mind, isn't in fact our conventional habitual grasping the more anomalous? So the question is: what vantage point are we looking from? (29)

> In the evening: deepening relaxation, more vividness. When space seems empty, watch carefully, pay closer attention, Alan says. I feel spacious: Oh, I see there's a lot more to observe. There's a feeling: here I'm really in unknown territory, never been here. No familiar ground. A sense of freedom of not being bound by beliefs, convictions, expectations. Then there's a sense of ambivalence and sadness, coming from who knows where. Followed by a sequence, a cascade, maybe in a few (milli) seconds:
>
> More sadness, and easily being irritated by sounds, cars coming and going, *anger* ...
>
> Comes in a feeling-flash as if it is done purposefully "against me," I'm a victim, I'm sought after … anxiety, a flash of *paranoia.*
>
> … And I feel micro-contraction, grasping at it: it gives a hold, this anger strengthens my self-sense!
>
> Then, the experience of multiple voices in me, one voice going in resistance: I don't want this, I can't stand it. More *anger.*
>
> Also seeing the futility of this agitation, the entropy energy,

the fuss; no way to go, a sense of being "cornered." Memories of Zen practice, feeling stuck with a koan, flash of desperate *rage.*

Experiencing powerless resistance shifting to sabotaging, ridiculing, "accusing the mindfulness" that is always right, sissy, silly. It feels unbearable that true mindful awareness invites me to truth and truthfulness!

Then, the *ridiculing, accusing* turns toward myself: not able to surrender, always doing things wrong ... Comes up: "I do things wrong, so I exist."

Then a shift to: feeling real *powerlessness*. Just feeling *sad*. And then a sense of great relief, joy, sense of freedom. Just this. (30)

Next: with the anger comes "I can't stand it. More anger." Powerless, narcissistic rage (rage coming from increased feeling of self-importance, in psychological view, or: coming from illusory separate-self sense, wisdom traditions).

Next: "I do things wrong, so I exist." I take space with that ... negative grandiosity (may be seen as counterpart to grandiose self experience: making and feeling oneself very important, not in a positive but in a negative way).

Nyam ... Today and earlier days, I hear "choruses" in the morning, forceful sounds! It feels threatening. I get the image: the building will be besieged. Fear, anxiety, we can be set on fire. A horn blowing, I hear some sort of text in it like: up to the fight.

This sense of threat – is there really something outside? What are you, blowing horn? Maybe some noisy workshop at the center? I go outside, and hear nothing. Are you in my room – some uneven noise in the heating system, now that today it's extra cold? Or is it just me, the noise something in me, unease, feeling unsafe? (31)

Next: This giving meaning, sort of projection of today's basic mood? A hallucination, must have been a perceptual distortion, with a world of meaning coming to it. So, next is a third person view to the earlier views.

I'm so aware of a growing acuity in fine-tuning of internal psychological processes and recognitions.

This morning I feel the start of a flush of indignation: some flavor of content of this arising object comes up, but the refinement is very much in the feeling tone, signaling, with a sense of early warning. When not present with it, I may jump over the feeling of anger. For instance, today at the table with breakfast stuff, another participant, to my idea, just cared for his own needs. He took all the (last) *raisins*, not leaving anything for us, for me. Indignation, and also, momentarily, there was a flash of a sense of neglect, a sense of lack. First, I experience:

A tendency to overreacting: inner and outer tendency of "acting out," in the inner sense with exploding; in the outer sense: almost grabbing that piled up plate, impulsively wanting to take a lot. With a flash of feeling that he looks angry at me.

Then, a flash of shame, then under-reacting: with a feeling of self-constriction, inner reactivity of suppression. (32)

Next: from suppression to dissociation (to be seen as a way of unconsciously defending against unwanted mental, emotional, behavioral processes by splitting these off from the rest of psychic activity). And then, able to access a larger awareness space again: feeling compassion, for him, for us, for me. As it went in reality that morning: after wordless questioning, looking at his face – then, with a smile taking one raisin from his plate ... gratefully receiving his smiling back. Explanation: "acting out" is the expression, in behavior, of still unconscious but emerging feelings; characterized by immediate action that bypasses the phase of trial action (thinking, feeling, experiencing).

Now that I've been here for some seven weeks, there's this extreme tiredness, a sense of paralysis, with pains throughout the whole body. Itching, a rash, feeling cold, "broken," shrinking, shriveling, needing ten hours' sleep, with a hot water bottle ...

I tell Alan, in the weekly interview: it must be something like experiencing "cold turkey," as I imagine that to be. Like waning from addictions, neurotransmitter shifts, body-mind feelings of withdrawal of chemicals. These swings in mood, sometimes almost maniacal and then down and out.

Much supine practice. Descending in tiredness of decades? Of lives? Trying to keep the mindfulness wide awake, while the rest is so sleepy. Mindfulness of sleepiness. Underlying it all, in some way surprisingly, in other ways not at all: this deep sense of joyful stillness. (33)

Time and space ... There is a sense of descending, like a rock, through flows and layers. Descending without descending. Then suddenly: a memory of thirty years ago comes up, somewhere on the beach, rain pouring down; being with a lover, exuberantly making love – now, excitement, indulging, sexual fantasies. Some energetic strain. Letting go of the grasping, still the fantasy arises, is, dissolves. No direct relation with what came before; things are stirring, the rock getting deeper, through sedimentation layers. Coming up: so much I've lived, so many images have ever been witnessed. There are memories, but so many have not been consciously processed, they've remained implicit. Now they come up, stuff just crawling out of the recesses and pockets, to be processed. From pleasures, light, and from dark side, unelaborated, desires, fears, pain – now giving a sense of unease, sadness.

"Descending," more and more into a vacuum, feeling some fear. Can I just stay with that, in an observing mode, not needing to get into anxiety of anxiety? This must be anxiety of "loosing self." Is this *horror vacui,* terror for emptiness? Feeling weightless, vulnerable in the body – is there a body? Surrender, quiet, relaxation. (34)

In the Daily Experiences Questionnaire and on *audio*: I describe that one day, end of the morning, the lunch car passed by. I heard it while meditating in my room, and a strong taste and clear visual image came up of something baked, light beige color with a golden-brown crust. When I went to eat, fifteen minutes later, indeed the meal was: minced meat with "beige" mozzarella, with a golden-brown crust from the oven. (35)

Remarkably, the same evening Alan tells about clairaudience and clairvoyance, the potential in anybody that is strengthened by shamatha practice. It's as if between the substrate of oneself and another, for a moment there can be a connection – through distance, through time. Seeing our lunch, in a flash, was this a wormhole experience? Maybe, a little inkling of something like that?

Here are some typical dreams from the second month:

> I have to find my way on a bicycle in the city to a particular address. Twice, the same thing happens: I know what it looks like, where I want to go (I think I know), but then I don't recognize the places on the way to going there. I have to look at the map; and then it turns out that this particular area is not represented on the map.
>
> Recurrent dream signs:
>
> Repeating theme, not knowing the way (while I thought I did), not recognizing, not having a good map (metaphor of some feelings, on retreat. There IS no "good map," there is no map – just surrender, live it!).
>
> Feeling lost, threatened in safety and integrity (and not seeing the dream aspect of it now), disorientation. (36)

> Boris and I are together with others, our intimate relationship has finished. We practice meditations related to the "Six Realms." I'm on a small sailing boat, exploring climbing the mast. At a certain moment the boat loses balance, tilts over and I fall into the water. It feels like a hard floor, concrete – pain, did I break my arm? Boris helps me out and then takes the same steps as I did before, in a sort of slow motion, to help me see what happened. I feel awake when we do this, I'm aware that I'm dreaming, and the

scene gets a joyful quality – this is fun! I feel love for him, warm gratitude. I want to climb this mast again, if this is a dream, I can't be hurt when I fall on concrete. I start climbing – and wake up from the dream.

Recurrent dream signs: water element. These months I've had so many dreams and imagery about being in water, an aquarium, sinking, letting myself float down (often through water-plants, back-bottom first). (37)

Comment: there is reference to the Six Realms of Existence, as presented in the Tibetan Buddhist "Wheel of Life," that shows the realms of hell, hungry ghosts, animals, jealous gods, gods, and humans. While traditionally taken as realms where a person may be reborn after death, they are also seen as psychological states in the human mind. More will be said about the Six Realms in Chapter 8.

Summarizing some impressions of turbulence, in this second month:

I've been experiencing emotions, feelings, sensations that seem to come up from deep recesses of the mind. In the physical sense: there has been a visible rash, with itching; there have been sensations of paralysis, shrinking, shriveling. And as to emotions, they have been strong, in the sense of fear, terror, anger, rage, with maniacal mood-swing flashes.

I see how, when there's new territory and openness, the reactive coarse personality-self in me takes its chance, in all possible creative guises and old overdue patterns. It's externalizing, projecting feelings to others, as a "you," with "What are you, blowing horn?" (31). Also, there's projection as an "it," with reification, making a threatening thing of a Zen teacher with koan study (30). There are perceptual distortions, hallucinations, there is grasping to new identities ("I do things wrong, so I exist"). And: next to the conventional perspective, I'm aware that also there is a continuity in observing from broader, more inclusive perspectives, with joy and a sense of knowingness. As to dreaming: there is lucid dreaming, or lucidity in parts of dreams, and a baby-step toward dream yoga.

Settling the mind: no preference for calm or active – guided meditation

After these diary sections about the first and second months, this is a remark to you, who are reading this book and have come this far. You have been offered many stories of bliss and turbulence ... are you still with me? It may be a good idea, for you – and me! – to do a guided meditation practice, before continuing to diaries of the third month. As Alan taught: when the mind is turbulent, surrender to Awareness of awareness: see for that practice Chapter 2.4. Alternatively, give yourself to Loving Kindness, as described in Chapter 3.2. Or join me, in another version of Settling the mind in its natural state, the practice that forms the basis for many experiences described in this chapter. This guided practice has been inspired by some themes that Wallace addressed during retreat, and in his book *Genuine Happiness*. It touches on our deeper motivation. I love his water element metaphor!

> Let us spend a few moments in discursive meditation, before we venture into the settling the mind practice, let's bring to mind our motivation for this session. In the context of your spiritual practice, and of your life in general: what do you most deeply and meaningfully yearn for? In what way could you find greater meaning, satisfaction, fulfillment, and happiness in your life? This aspiration may pertain not only to your individual well-being but may include all persons with whom you come in contact, and the world at large. It is helpful to bring to mind your most meaningful motivation; this will enrich your practice.
>
> We begin with stabilizing this marvelous instrument, the mind, with refining its focus of attention. Place your attention once again upon the tactile sensations created when the air enters and exits your nostrils. And, with introspective awareness to the state of your mind, see whether it is agitated, drowsy, calm, excited, or alert ...
>
> We have six doors of perception, including the five physical

senses of sight, sound, touch, smell, and taste. Next to these, we have another avenue, a direct perception into another realm of reality – the sixth domain of mental experience. You may take an interest now in the domain of the mind, this field of experience in which thoughts, feelings, images, emotions, memories, plans, desires, fears, and at night, dreams, all occur. For the time being, for now, we will also maintain a peripheral mindfulness of the breathing. In a metaphorical way, let's keep one hand of the attention on the buoy of the breath. In this metaphor, imagine you are out in the ocean in a gentle swell with rising and falling of waves; you keep one hand on a buoy to give you some stability, and point of reference.

(*Wallace further elaborates on the metaphor*) In this imagination, wearing a facemask, you sink your head beneath the surface of the waves, aware with your body of the rise and fall of the incoming swells. At the same time you attend to whatever lies in the ocean depths, in the transparent, luminous water.

You observe whatever comes up in the mind with no preference as to whether thoughts are arising or not, and no preference for nice thoughts as opposed harsh thoughts. You observe, without intervention or manipulation, without control. Let the thought rise up, pass before you, then fade away. Practicing in this way, the mind gradually settles in its *natural* state, which is quite distinct from its *habitual* state. Habitually, the mind oscillates between excitation and laxity. Yet, with this practice the mind gradually comes to rest in its ground state, that is calm and clear ...

While in this practice there is no preference for having a calm and quiet mind as opposed to an active and discursive mind, it *is* appropriate to prefer to not be distracted. It is appropriate to not be carried away by the events of the mind and not grasping onto them. Is it possible for you to be calm even when the thoughts passing through your mind are turbulent?

Continue breathing into the abdomen as usual. If you become

disoriented, return to the breath in the abdomen, stabilize a bit, and then hover in the space of the mind once again …

To bring this session to a close, dedicate whatever has been meaningful in this practice, to the fulfillment of your aspirations for your own well-being and happiness and for those around you. May this practice be of great benefit.[1]

5.4 Third month meditation practice

Now, let's continue with our "n=1 case study empirical research" project for the third month. Object and subject in this phase often do not present in as distinct ways as before; they are taken together in the presentation of the diary fragments.

> Mind, Settling … Often coming up, as objects: feelings and insights about the practice, the intelligence in it, in refining attention. So great refining still to go! The practice: not taking anything away, or adding anything from outside. Continuity and increasing completeness. More wide space in head and mind, all-inclusiveness. Feels like opening up potentials that were already there, have always been – while the practice itself is creating new wholesome habits, in the sense of non-grasping, with greater presence. (38)

> The beauty of this stillness and movement, mind resting, events coming up. Increasing familiarity. A feeling of trust, non-referential security, of holding, and being held. These different dimensions of awareness … Practically, the Settling the mind meditation setup offers sort of a safety valve, the possibility to vary, to quieten the process a bit, by shifting the emphasis for some time to space of mind, to the background. Being with objects arising, or: just objects arising in awareness, with awareness as the vantage point.
>
> The union in stillness and movement: space and content are inseparable. Awareness "holding" the observer, until the observer

merges in awareness. (39)

Comment: "holding" in psychodynamic therapy terminology refers to a supportive function, in the sense of the therapist's holding in mind, and acting as a secure container for the fears, guilt feelings, pains and turbulences in the mind of the patient. The sense of "holding" referred to here has a nondual taste.

While resting, awareness in awareness, again and again ... it feels like a shift, all along: the "host" manifesting as witnessing, and then as knowingness, cognizance. All along: the path being the fruit ... (40)

Some examples concerning awareness and monitoring, in the third month:

Continuity, completeness, going by themselves, taking over.

This openness, the correlates in whole brain, including "right hemisphere," relating to patterns and configurations, to space. Patterns of light, energy. With this diary I feel the need to make drawings for representation, more dimensional, feeling that two dimensional text is too limited. Glimpses in bliss, clarity-luminosity, cognizance, glimpses in knowing things to their nature. Now, without conceptual overlays ... thus-ness shining through, as it is. The trust, that this knowing is always available. I'm writing less and less. (41)

Monitoring evolving to more and more subtle, there is awareness, it feels like an "infinite regress" of awareness of awareness; just being that. No viewer, no viewing, no viewed. No center, no periphery: a non-local awareness – with "me" distributing in space, dissolving. (42)

This on-going flow of knowing, session after session after session. This "being knowing," timeless, still. A joyful silent

> beaming kind of excitement, gratitude. Sometimes it feels like too much for the organism. So radiantly clear, so razor-sharp ... moving to tears. (43)

In one of the *audio* recordings, I express my amazement about the way that every day shows a different kind of landscape, or mindscape; that actually every session of Settling the mind has a different taste, a different atmosphere. (44)

In the *video* interview I make remarks, with a concrete example, about a subtle shift in consciousness that I've been acutely aware of on a specific moment. There was the experiencing of a "receiving" aspect, as like in radar, or sonar – in some sort of a local way; yet, also the experiencing of being a non-local receptacle of knowledge that is "all around in" consciousness, that *is* consciousness – it's both. The glimpses "I am this, I am this knowledge, this cognizance." (45)

Looking back at these descriptions regarding the third month: the grasping seems again to be lessening, and becoming more and more subtle; there is more space and calm. The practice: just this ... Arisings and dissolutions are more brief and passing. Subjectivity is more often resting in space, and more often dissolving in space. As there is less of objects, there is less and less of a subject. There are the glimpses in knowing that both objects and subjectivity are facets of awareness itself. Regarding the space: there is more experience of a sense of cognizance, awareness, being-ness. And also, more insight in the psyche, more discovery aspects are coming in, that combine with the development aspect.

There is a sense that I had more and more difficulty finding any stammering words and vocabulary for what I experienced; coinciding with less inclination to write. And all the time then, and now, it's so evident that there is so much more refining to go.

Meditation turbulence, third month: "yo-yoing" and "whiplash"

Here are some instances of turbulence and *nyam* experiences during meditation practice, for this third last month in the retreat:

> Non-distraction and non-grasping seem to facilitate some movement of falling out of habitual conditioning. The falling happens so often, more and more understood as a gradual *"yo-yoing"* movement, in a way that it can be integrated: that's what I mostly experience, as a "witness in wonderland." Sometimes the falling seems to go too quick for the organism to be integrated, there's a sudden, unexpected shock ... and there is a "finding back of myself" in a subtly tense, contracted state. When too anxiety provoking, there is a sense of a micro counter-coup, like a reflex *"whiplash"* of contraction, outside of my control. (46)
>
> *Next: A feeling that all falling out of conditioning, all falling-losing, giving away control, is, basically, "toward the substrate." And in the process of yo-yoing the idea of gravity dissolves; just non-local, awareness. Whiplash of contraction: feeling as a sudden involuntary body-mind reflex.*

I'm so aware of the way that an emotional "grasping-issue," even a subtle feeling of irritation or anger, can suddenly fill the complete space. The color often feels dark red-brown. Today there was this movement: a sudden exploding and expanding into a larger space, the whole universe – all-filling "righteous indignation," anger, a flash of grandiosity, and then the whole universe feels "me." There's seductive power in this. Next micro-moment, it's rather as if the space of mind is contracting, mind imploding to small, as if forming a membrane, closing off, locking up experience. When both energy-emotion movements are acknowledged, there is relaxation and space (47)

Next: "Ignorance, reinforcing the ego-personality." Before the contracting: there is depersonalization (a sense of being unreal). Comment: in relation to space of mind, at a later moment Wallace addressed comparable experiences, in a way that struck me. He said that our minds appear to become small, when we get fixated on something. Trivial issues may seem to become large and important, in our awareness. It appears that we experience the magnitude of the contents of the mind as relative to the spaciousness of the mind.

Some *nyam* of the last few days:

Visual: on the ceiling moving forms, light and dark, a little black spot that takes my attention when I'm lying supine, preparing for a vacuous gaze. It looks like a small insect, seems moving, creeping. Is it just the liveliness of color patterns that I connect with some familiar notion? The same with some other spots, on wood. (48)

In the third *video* interview, I share the feeling of being impressed with how, in a global way, Western psychological approach mostly takes such a limited band width of consciousness as the "normal habitual" life. A psychological approach with such remarkable achievements; yet, there is this limiting normalcy view. The view that there are, on the one hand (I make movement "up" with my hand) persons with problems, with psychosis, borderline problems, neurosis. And there is normality, to which one may return. Mostly there is little conceiving (movement "down"), of a possible perspective of exceptional health, and how this can be won. This limiting band width of so-called normalcy seems to culturally confirm that "this is what exists," nothing outside it. And then, the training in attention in this project shows that there can be much reinforcement from wider dimensions of awareness. There can be more pliancy, more flourishing. As I say in the interview: this has been so important for me, the path being the fruit: this self-reinforcing joy and happiness, that can't be preconceived, can't be predicted or imagined, that really "feels true." (49)

A typical dream during the third month is the following:

I am on retreat. I hear that maybe the complete staff will be away for one week. This, I believe, will be impractical now that we've been informed that the language is going to be changed completely, which will require many adaptations, give a lot of extra work (change signs, announcements, notices and so on, also a lot we don't know yet), and we don't have much spare time. Feeling: fear to be left alone, some agitation, worry.

Anomalous dream sign: complete change of language would be rather odd … still metaphorically makes sense, in this retreat. Willed disorientation and re-orientation? (50)

Yes, there is some willed disorientation, when one chooses to go into daylong meditations. This can't miss bringing the body-mind into new dynamics, new states of being. We participants, in a way, have been actively looking for that. Still, it arouses fear, of course. During daytime I'm often more aware of the excitement that goes with the openness, the spaciousness. Yet, in *nyam* and dreams at night some fears manifest and translate into storylines, using some impressions of the day. The dreams invite for lucidity, for falsification of imagined threats by seeing a dream for what it is. But when I'm not lucid while dreaming, like in this one, I'm identified with this worried separate-self sense, and this agitated body-mind.

Following are some more impressions of turbulence in the third month. There is increasing openness and generally less fear. I'm more aware of the felt sense of *yo-yoing* and reflex *whiplash* contraction dynamics – these have been the names that came up. While the yo-yoing felt sense is a more gradual dynamic, the whiplash became more conscious in this third month: maybe with more sudden falling through, more edgy micro-movements, more turbulence that aroused reflex defense by the body-mind. Additionally, increasingly subtle sensitivity will play its part. Still, there is also increasing speed of recovery and integration, there is increasing flow and lightness. More about yo-yoing (24, 46) and reflex whiplash (46) will be said in Chapter 9.

"Witness in wonderland," yes, at moments, it feels like that. With

Settling the mind practice as a serious yet joyful inner playground for exploration of the psyche.

5.5 Just sitting

When I look back over these three months, next to the increasing continuity and completeness in focus, I notice more and more refining and melting of object, subject, and ways of monitoring and awareness. When part of me falls back into conditioned ways of defending and coping with anxieties, the quicker recovery is "re-covery" in a special sense: it is not a coming back to an old state. It is a new balance in a new way of being. Being present with what is, not being identified with turmoil, neither with defenses against turmoil.

The widening scope and the positive reinforcement of "the path being the fruit" brought up in me the wish: even if we would look at it from a (in this respect limited) psychotherapeutic perspective that doesn't transcend identification with the separate-self sense: also then a brief period of guidance in being with whatever comes up will be so helpful. For anyone, some practice in non-grasping, non-identification, and getting a sense of this positive reinforcement would be so nourishing! With the upheavals and turbulences, still, there is this underlying sense of joy and happiness, touching on a sense of familiarity and at homeness. There is such a wholesome quality in experiencing more "degrees of freedom" in having the courage of just seeing what is, without judging, without needing to arm, defend, idealize, distort, and build walls. Interestingly, for part of me the process initially coincided with a sense of relief of "surviving," and sure you do survive, even if it's not who you think is you!

Experiencing new inner territory ... The turbulences that I describe with each month, including fear, a sense of paranoia, derealization, obsessional thinking, in some sense connect different fields of interest and involvement in my life. These are meditation experiences – and these are experiences that I have been familiar with, sometimes not so much in a personal way, as with others. With patients, clients who came to me for psychotherapy, I've seen how helpful a "just being with," sitting-through process could be. I'm thinking now of Emily, a young slender woman painter, who sought therapy because of problems in relationships. What struck me at our first meeting was the tense smile on her face, almost a grin. She was often tapping with her fingers on her leg. Her style of coping with anxiety, agitation and feelings of unsafety, we got to see, had always been to try to escape, to not feel, not think. She sought to not be aware of inner turmoil, but to create more external turmoil in order to seemingly hide from the internal ordeal. When in stress and fear she would start a new impossible relationship with "exactly the wrong man," rather than stop and be present – as she started to see. While never using the word meditation or mindfulness, just the stopping and being present with what is, being able to observe without needing to act, internally and externally, has been such a help for her. The pain is in the grasping, in the identification with the separate-self sense and experiences. The grasping identification can manifest in many tastes of – to phrase it in Buddhist notions – attraction (I need this fuss, dust, drama, this man) and aversion (I can't stand that silence, sadness). In a way, this grasping stands for lack of trust. Trust in who we are.

While not easy all the time, I feel so grateful for having been able to explore unknown ranges in the mental landscape, in the relatively safe environment of retreat. And in some sense I feel more connected with clients who have given me their trust in sharing their paths, turbulences included, in our psychotherapy sessions. One mind …

Chapter 6

Settling the mind and "tasting" the texts

Acting as your own mentor, if you can bring the essential points to perfection, as if you were threading a needle, the afflictions of your own mind-stream will subside ... (Lerab Lingpa, 6.1)

In this chapter I present some excerpts of relevant texts by nineteenth century Dzogchen masters Lerab Lingpa and Düdjom Lingpa, specifically on the Settling the mind practice. These are the texts that Alan Wallace referred to during the Shamatha expedition on a regular basis. Lerab Lingpa's text presents the basics for the practice, while Düdjom Lingpa elaborates on special experiences one may have when doing this meditation.

For me, after having practiced for three months, with experiences as described in Chapter 5, studying these texts later has felt like zooming in on many intricate details, allowing for more and deeper understandings and recognition. This must be the recognition of what finding oneself in completely new mind territory may bring in. Referring to "tasting the texts" in the chapter-title may connect with the surprising synesthesia (perception of stimuli in various sense modalities at the same time, as described in Chapter 5.2): next to seeing, hearing, and mentalizing, there must certainly also be a touch, smell, taste correlate for this perceptual process!

Lerab Lingpa (1856-1926) was a great master, a famed Tertön (discoverer of spiritual treasures) and the teacher of the 13th Dalai Lama. He found many spiritual treasures of Padmasambhava. Alan Wallace told us that he teaches Settling the mind in its natural state practice in the way as described by Lerab Lingpa in his *Commentary to the Dzogchen teachings called Heart Essence of Vimalamitra*. Vimalamitra, like Padmasambhava, was an eighth century Indian Buddhist master, who later taught in Tibet.

While Düdjom Lingpa (1835-1904) was not a monk, being married and having eight sons, he performed many miracles and allegedly reached the highest levels of realization, in Tantra as well as the Great Perfection. It is said that he received *The Vajra Essence* teachings – from which I will quote – from the dharmakaya, the ultimate ground of reality, the Buddha mind. In Wallace's words: the *Vajra Essence* was essentially "downloaded" from the dharmakaya. *Vajra* stands for both thunderbolt and diamond. The *vajra* is a ritual object that symbolizes the properties of both: irresistible force and indestructibility, respectively. After reception of the teachings by Düdjom Lingpa, it took around thirteen years before they were manifested in the world and made public. Interestingly, in one of Düdjom Lingpa's dreams it was prophesized to him that the benefit of his profound hidden treasures would go West, with the words "Those deserving to be tamed by you dwell in human cities to the West."[1] So, it may have been meant to be read by us Western barbarians, with over-stimulated, un-tamed minds.

Reading these texts, I sometimes feel excited and in awe. The text by Lerab Lingpa impresses me with its timeless, down-to-earth reminders. Düdjom Lingpa's exposition of mind-body turbulences during practice, about what he names "signs of progress" I find thoroughly intriguing.

In this chapter, after presenting a text, I give some comments. Then I place a few sections and items of the Buddhist experiential texts next to some – surprisingly – corresponding diary items as have been presented in Chapter 5.

6.1 Basic instructions by Dzogchen master Lerab Lingpa

The Settling the mind in its natural state meditations, in the guidance by Alan Wallace, as presented in Chapter 2.3 and other chapters, are based on the following instructions. I invite you to apply this reading as a guided meditation for yourself, while remembering that thousands of meditators in the nineteenth, twentieth and twenty-first centuries, in and outside of Tibet, have followed the same instructions.

Lerab Lingpa's Settling the mind – guided meditation

After some introductory remarks, Lerab Lingpa offers these guidelines, part of which are presented here in paraphrasing.

"Simply hearing your spiritual mentor's practical instructions and knowing how to explain them to others does not liberate your own mind-stream, so you must meditate. Even if you spend your whole life practicing a mere semblance of meditation – meditating in a stupor, cluttering the mind with fantasies, and taking many breaks during your sessions due to being unable to control mental scattering – no good experiences or realizations will arise." After mentioning these considerations, Lerab Lingpa invites you to practice in solitude, sitting upright on a comfortable cushion. "Gently hold the 'vase breath' until the vital energies converge naturally. Let your gaze be vacant. With your body and mind inwardly relaxed, and without allowing the continuum of your consciousness to fade from a state of limpidity and vivid clarity, sustain it naturally and radiantly." He advises you to not clutter your mind with many criticisms and judgments, and to not take a short-sighted view of meditation. It is better to avoid great hopes and fears that your meditation will turn out one way and not another. Lerab Lingpa suggests that, in the beginning, you have many daily sessions, each of them of brief duration, and to focus with great precision. His core advice is, whenever you meditate, to bear in mind the phrase "without distraction and without grasping," and to put this into practice.

Then, as you gradually familiarize yourself with the meditation, you can increase the duration of your sessions. If dullness comes in, you arouse your awareness. With excessive scattering and excitation, you loosen up. He advises you to determine in terms of your own experience the optimal degree of mental arousal, as well as the healthiest diet and behavior.

After explaining these basics, Lerab Lingpa continues:

"Excessive, imprisoning constriction of the mind, loss of clarity due to lassitude, and excessive relaxation resulting in involuntary vocalization and eye-movement are to be avoided. It does only harm to talk a lot about such things as extrasensory perception and miscellaneous dreams or to claim, 'I saw a deity. I saw a ghost. I know this. I've realized that ...' and so on." He notes that the presence or absence of any kind of pleasure or displeasure, such as a sensation of motion, is not uniform, and that there are great differences in the dispositions and faculties from one individual to another. Then, due to maintaining the mind in its natural state, there may arise sensations such as mental and physical well-being, a sense of lucid consciousness, the appearance of empty forms, and a nonconceptual sense that nothing can harm the mind, regardless of whether or not thoughts have ceased. As he states: "Whatever kinds of mental imagery occur – be they gentle or violent, subtle or coarse, of long or short duration, strong or weak, good or bad – observe their nature, and avoid any obsessive evaluation of them as being one thing and not another." Lerab Lingpa invites you to let the heart of your practice be consciousness in its natural state, limpid and vivid. "Acting as your own mentor, if you can bring the essential points to perfection, as if you were threading a needle, the afflictions of your own mind-stream will subside ... This is a sound basis for the arising of all states of meditative concentration on the stages of generation and completion."[2]

Some comments to the text by Lerab Lingpa

Lerab Lingpa writes and addresses you in a sober and down-to-earth way, as if he's sitting in front of you. Well, he wasn't really, but for me he often came in, as if channeled by Alan Wallace, sitting only a few meters distance from where I daily "abided" on the cushion. Reading these words I don't just "hear" Lerab Lingpa's voice in an imaginary way, but I also hear Alan's voice, in a memory. Here are some of Wallace's clarifications to this text, that he gave during the Shamatha Project retreat. Next to that, some other teachers' voices are added, and a few of my musings.

Lerab Lingpa refers to gently holding the "vase breath" until the vital energies converge naturally. When you practice what is called "gentle vase breathing," as you inhale you let the sensations of the breath flow down to the bottom of the abdomen, like pouring water into a vase. When you breathe out, you retain some air in the lower abdomen. According to Wallace, this contributes to energizing and stabilizing breathing. While supporting the convergence of the vital energies, or *prana*s, in the central channel in your abdomen, allow them to settle in this region. Thrangu Rinpoche advises the practice of gentle vase breathing in shamatha breath meditation for supporting clearing our mind if it is unclear, and stabilizing it if it is unstable. [3]

Regarding the words "Let your gaze be vacant": as was clear in Wallace's meditation instructions, the eyes are open in this practice, with vacantly resting one's gaze in the space in front. By leaving the eyes open, while focusing on the domain of mental events, the artificial barrier between inner and outer begins to melt. It seems, we naturally have a sense that we are looking out on the world from behind our eyes. However, the sense of an independent subject inside the head is an illusion. Wallace stated that there is no scientific support for such a belief. With inspection through rigorous contemplative inquiry, an autonomous thinker and observer inside the head has not been found. What we know is that mental events are correlated with neural events – which doesn't necessarily mean that their location is in the same place. Thrangu Rinpoche quotes the great master Saraha, with whom in the eighth century an important Mahamudra-lineage allegedly started. Saraha states that not-closing the eyes means that the meditator, wide-eyed, without even blinking, meditates with *Vajra* eyes. Regarding practitioners in the Vajrayana tradition it is said that: "*Vajra* eyes looking directly" means that these practitioners look neither up nor down; they simply look straight ahead. What exactly are they looking at? They are looking at their own mind. [4]

As the text continues: " ... without allowing the continuum of your consciousness to fade from a state of limpidity and vivid clarity, sustain your awareness naturally and radiantly." The limpidity of awareness refers to the qualities of transparency and luminosity, qualities of awareness itself. This practice, next to its developmental aspect, is one of discovering the innate stillness and vividness of awareness.

Diary October 2007

> To my feeling, initially we cultivate the developing aspect in this ever refining training especially with the "moving"; and with the "stillness" we get to feeling more and more at home with the space of mind, and with discovery. In their complementarity there is also a sense of discovery – with, in the end, their "sameness," being one ... as has always been.

An optimal degree of mental arousal

"Determine in terms of your own experience the optimal degree of mental arousal, as well as the healthiest diet and behavior." Here is the question of how to find this optimal degree, free of dullness and excitation – and how to deal with possible imbalances. As Alan often stated: the primary challenge is to overcome laxity without undermining stability. In his view, you counteract laxity by arousing the attention, taking a greater, more intense interest in the object of meditation. However, if you arouse the mind too much in your efforts to remedy laxity, it will easily fall into excitation. And if you relax too much, you will likely succumb to laxity. It's a delicate balancing act. You can only meet this challenge by way of your own experience, sensing for yourself the suitable degree of effort to tune your attention. Taking a greater interest with greater arousal, and relaxing, and finding the balance. Other teachers use different terms, and go into varying detail about skills in this context. As to terminology, to name a few variations: Dakpo Tashi Namgyal speaks of the vital point of balance between exertion and relaxation, Traleg

Kyabgon Rinpoche speaks of tightening and loosening, and Daniel Brown refers to intensifying and easing up. [5]

Regarding the healthiest diet and behavior: as to diet there may have been many tastes through the ages. As to behavior: certainly the yoga sessions that one of our participants offered have been very helpful for me. Now, after the project, I'm so aware that shamatha – and flourishing health in general – does include also the physical aspects. In the present time I'm gratefully enjoying the *Kum Nye*, Tibetan yoga approach, developed in forms accessible for Westerners by Tarthang Tulku Rinpoche, that views the body and mind holistically, integrating and balancing the physical, the psychological and the spiritual for greater health.

Lerab Lingpa admonishes us that excessive relaxation resulting in involuntary vocalization and eye-movement are to be avoided. He adds that it is only harming to talk a lot about such things as extrasensory perception and miscellaneous dreams. Later, he mentions: "The presence or absence of any kind of pleasure or displeasure, such as a sensation of motion, is not uniform, for there are great differences in the dispositions and faculties from one individual to another." Wallace mentioned an old Tibetan saying: "If you fill a gourd with just a little water and shake it, it makes a lot of noise. But if you fill it to the brim and shake it, it makes no sound." Making any claims about one's meditative achievements – even if they are true – creates obstacles to one's practice, thus was found, through generations of contemplatives. Special experiences and surprising perceptions, pleasant and unpleasant, in principle are just to be shared in private with one's spiritual teacher. Well, we on Shamatha expedition were in a way protected for that, by the silent set-up of the expedition. While I get the value of not making claims about one's contemplative achievements, I have also found sharing my experiences with teachers and peers to be very valuable. There is a difference between boasting, and comparing notes and learning from each other.

Tilling the soil of a field

"Due to maintaining the mind in its natural state, there may arise sensations such as physical and mental well-being, lucid consciousness, the appearance of empty forms, and a nonconceptual sense that nothing can harm the mind, regardless of whether or not thoughts have ceased." As Alan noted: all that appears to the mind is an interrelated matrix of sensory phenomena, but these phenomena and qualities no longer appear to belong to something that is absolutely objective, as the sense of reified duality is diminishing. Empty forms: even your own body appears empty of substance. So, here the notion of emptiness, so paramount for vipashyana-insight practice, is included: phenomena being empty of inherent existence, as they are interconnected, non-separate and insubstantial. Regarding the mind: when there is less and less distraction and grasping, one simply observes thoughts and images arising in the present, without going in to referents, meanings, projections, without being either attracted to them or repulsed by them or anxious of them. Yes, this is the training that I described in Chapter 5: being with the turbulences, neither attracted to these or repulsed by them, or anxious of them. Then, as Lerab Lingpa states, the afflictions of your own mind-stream will subside. There is increasing emotional balance and equanimity. Lerab Lingpa compared this way of practicing with tilling the soil of a field: you establish a foundation that is supportive for whatever further practice you will perform. I like the way Alan Wallace refers to what happens: one's psychological immune system is strengthened by this practice, so when events occur that were previously upsetting, one can now cope with them with greater composure. One has a heightened sanity. [6]

Diary after retreat, 2010

> "It's just a thought!" I remember Ellis, a Zen priest friend saying this to me long ago, while taking a bite of her cookie, going with her green tea. We were sitting on a bench, outside, at the Zen

Center where she lived and did her training. It was early spring, subtle green promise transpiring in the trees. In a sense, she may have wanted to reassure some anxious part in me, at that time. In that sense, she meant: it can't harm, it is not what is called "real" material reality, touchable, and it's not independent. Still, then, I felt a bit helpless: yes, it's a thought, it is my thought, and it's important, it expresses something that's real to me, that influences me. Indeed, if I took separate-self sense very seriously, if I took this thought to be "me," I could be threatened, in danger – the more, I would say now, if things are repressed and dissociated.

Now I'm thinking back of that sense of "danger in the space of mind." The danger, in this context not so much being outside, in what's called real life, but in unwholesome thought-conditionings, in grasping. Thoughts have their own reality, but the imaginary threats, against which "self" needs to be protected, are very much confused-mind made. And now I'm aware that there's more trust in me, also in the sense that I like to take more risks, to practice in this refined, honest way with myself, and also with others.

Regarding the stages of generation and completion: these are the two phases of Buddhist tantric practice. In Chapter 4.1 a brief allusion was made to these stages, in referring to visualization of enlightened forms (deities and mandalas), and cultivation of the subtle energies of the psycho-physical body, along with recognition of our Buddha nature. Good achievement in shamatha will offer a strong foundation for all advanced practices.

6.2 Dzogchen master Düdjom Lingpa: *The Vajra Essence*

While the turbulence aspect fascinates me, connecting with my background in psychiatry, it also is something that often plays a role in practicing communities, but doesn't get so much attention.

Düdjom Lingpa, in *The Vajra Essence,* presents a unique and captivating account of the practice of Settling the mind in its natural state, with a remarkable section that describes rather turbulent happenings in his students. He addresses them as "signs of progress." Settling the mind shamatha practice is presented by Düdjom Lingpa as: "taking appearances and awareness as the path." His *Vajra Essence* figures as a classic nineteenth century meditation manual in the Dzogchen tradition. Alan Wallace has translated this text, with the guidance of Gyatrul Rinpoche. Clearly, Wallace is a great adept to both the text and the Settling the mind practice. Of all the shamatha practices known to him, Wallace says – and he practiced dozens of them – he finds none more intrinsically interesting than this one. According to him two marvelous things take place. First, like with the Delphi oracle, this is about getting to know your mind. You have a front-row seat, welcome to yourself! And second, this is a profoundly therapeutic process, in watching the innate healing capacity of your mind, seeing the unraveling happening.

Often, Buddhist meditation manuals present accounts of what to expect in the practice when it is done in the right way. They tell what should happen, not what often does happen even when practice is done correctly. Next to that, *The Vajra Essence* also tells of a wide array of "meditative experiences," or *nyam* in Tibetan, that may occur in shamatha meditation, and specifically with Settling the mind in its natural state. Düdjom Lingpa's instructions for Settling the mind are as follows. " 'Now to remain for a long time in the domain of the essential nature of the mind, I shall be watchful, observing motion, keeping my body straight, and maintaining vigilant mindfulness.' When you say that and practice it, fluctuating thoughts do not cease, but without getting lost in them as usual, mindful awareness exposes them. By applying yourself to this practice constantly at all times, both during and between meditation sessions, eventually all coarse and subtle thoughts will be calmed in the empty expanse of the essential nature of your mind." As to what you as a meditator may experience, Düdjom Lingpa adds: "You will become still in an unfluctuating state, in which you experience joy like the warmth of a fire, clarity like the dawn, and nonconceptuality like an ocean unmoved by waves. Yearning for this and believing in it you will not be able to bear being separated from it, and you will hold fast to it ... That is called ordinary quiescence of the path, and if you achieve stability in it for a long time, you will have achieved the critical feature of stability in your mindstream." [7]

So, this is about the stillness of the space of mind, and the motion of thought. Düdjom Lingpa speaks of applying yourself to this practice both during and between meditation sessions: indeed we need the formal meditation sessions, however, in-between we can take on the attitude and awareness of this non-grasping practice. Regarding his words that you will not be able to bear being separated from it, and you will hold fast to it: not knowing any better, you might think that this is it! But as we saw, there is no full liberation in shamatha. It's the foundation for further explorations.

Signs of progress

Experiences differ greatly from person to person, as everyone's mind is so utterly complex. The following wide-ranging collection of meditative experiences, as described by Düdjom Lingpa, may arise during the Settling the mind training, especially when it is pursued in solitude for many hours a day, for many months. We in the Shamatha Project may have experienced a little trailer of this movie!

Düdjom Lingpa names the following phenomena as signs of progress for individuals who take appearances and awareness as the path. He refers to:

(1) The impression that all your thoughts are wreaking havoc in your body, speech, and mind, like boulders rolling down a steep mountain, crushing and destroying everything in their path.

(2) A sharp pain in your heart as a result of all your thoughts, as if you had been pierced with the tip of a weapon.

(3) The ecstatic, blissful sense that mental stillness is pleasurable,

but movement is painful.

In *(4)* Düdjom Lingpa notes that a meditator may perceive all phenomena as brilliant, colored particles.

He continues with *(5)*: Intolerable pain throughout the body from the tips of the hair on your head down to the tips of your toenails.

(6) The sense that even food and drink are harmful as a result of being tormented by a variety of the four hundred and four types of identifiable, complex disorders of wind, bile, and phlegm.

(7) An inexplicable sense of paranoia about meeting other people, visiting their homes, or being in town.

(8) Compulsive hope in medical treatment, divinations, and astrology.

(9) Such unbearable misery that you think your heart will burst.

(10) Insomnia at night, or fitful sleep like that of someone who is critically ill.

(11) Grief and disorientation when you wake up, like a camel that has lost her beloved calf.

(12) The conviction that there is still some decisive understanding or knowledge you must have, and yearning for it like a thirsty person longing for water.

A thirsty person longing for water

... Well, for a little pause here: these are strong emotions! Deep longing and yearning are described. Düdjom Lingpa continues, referring to additional phenomena that a meditator may experience:

(13) The emergence, one after another, of all kinds of thoughts stemming from the mental afflictions of the five poisons, so that you must pursue them, as painful as this may be.

(14) Various speech impediments and respiratory ailments.

As Düdjom Lingpa mentions *(15)*: all kind of experiences can occur – called experiences because all thoughts are expressions of the mind, where all appearances of joys and sorrows are experienced as such and cannot be articulated – yet all experiences of joys and sorrows are simultaneously forgotten and vanish.

(16) The conviction that there is some special meaning in every external sound that you hear and form that you see. You may think: "That must be a sign or omen for me"; and compulsively speculating about the chirping of birds and everything else you see and feel.

In *(17)* Düdjom Lingpa refers to meditators having the sensation that external sounds and voices of humans, dogs, and birds, are piercing their hearts like thorns.

(18) Unbearable anger due to the paranoid thoughts that everyone is gossiping about you and disparaging you.

(19) Negative reactions when you hear and see others joking around and laughing, thinking that they are making fun of you, and retaliating verbally.

Düdjom Lingpa, in *(20)*, points at meditators who in a compulsive way are longing for others' happiness when they watch them, due to their own experience of suffering.

(21) Fear and terror about weapons and even your own friends, because your mind is filled with a constant stream of anxieties.

(22) Everything around you leading to all kinds of hopes and fears.

(23) Premonitions of others who will come the next day, when you get into bed at night.

Hopes and fears

From time to time, take a conscious in-breath and certainly also a deep out-breath, while reading. In addition to the named intense experiences Düdjom Lingpa refers to the following:

(24) Uncontrollable fear, anger, obsessive attachment, and hatred when images arise, seeing others' faces, forms, minds, and conversations, as well as demons and so forth, preventing you from falling asleep.

(25) Weeping out of reverence and devotion to your gurus, or out of your faith and devotion in the Three Jewels, your sense of renunciation and disillusionment with samsara, and your heartfelt compassion for sentient beings.

(26) The vanishing of all your suffering and the saturation of your mind with radiant clarity and ecstasy, like pristine space, although such radiant clarity may be preceded by rough experiences.

(27) The feeling that gods or demons are actually carrying away your head, limbs, and vital organs, leaving behind only a vapor trail, or merely having the sensation of this happening, or experiencing it occurring in a dream. Afterward, all your anguish vanishes, and you experience a sense of ecstasy as if the sky had become free of clouds. In the midst of this, the four kinds of mindfulness and various pleasant and harsh sensations may occur.

In *(28)* Düdjom Lingpa gives advice to those who guide others on this path: spiritual friends who teach this path properly must know and realize that these experiences are not the same for everyone, so bear this in mind!

For conclusion of this overview, Düdjom Lingpa addresses the meditator, saying that after all pleasant and harsh sensations have disappeared into the space of awareness by just letting thoughts be, without having to do anything with them – all appearances lose their capacity to help or harm, and that you can remain in that state. He mentions that you may also have an extraordinary sense of bliss, luminosity and nonconceptuality ... and a small degree of extrasensory perception *(29)*. [8]

Some comments to the text by Düdjom Lingpa

For me, hearing about many of these "signs of progress" as Alan mentioned them during retreat, and then reading this long list for the first time in one swallow gave some spacious excitement, and sort of breathlessness. This may have been a *nyam* by itself, catalyzed by the hearing and reading-meditation alone! The great variety in what may be experienced, the expanded mental landscape ... A mindscape, painted in colors and meanings of: this can happen, don't worry too much, these are signs of progress, in the context! Still, maybe I should have told you in advance to "fasten your seatbelts," as I was told to do, in a dream (mentioned in Chapter 3.5).

Wallace expressed to us his feelings of wonder about the named phenomena, and notes that many of us would likely respond to some of those unpleasant experiences by dropping out, and maybe seeking medical help. And Düdjom Lingpa calls them "signs of progress"! Certainly, it is progress when we recognize how turbulent our minds are. Wallace adds: " … the deeper you venture into the inner wilderness of the mind, the more you encounter all kinds of unexpected and, at times, deeply troubling memories and impulses that manifest both psychologically and physically." The deepening mind-body relaxation can lead to strange experiences, like: the body may show muscle shocks and myoclonic jerks that come up when you are really relaxed. These are mind-body correlates, showing in the *prana* system. As we learnt, in general, with *nyam*, just "watch your mind heal!" Keep observing, with no distraction, no grasping. According to Wallace, the only way to probe the depths of consciousness is going by way of the psyche, with all its imbalances and neuroses. [9] Whatever comes up was already there, previously concealed by the dullness and turbidness of the mind.

How can one conceive of these intense experiences? A few remarks are given here and more will follow. If anything, this variety of experiences may cover a vast phenomenological overview of physical and mental health experiences.

Dragons and demons

Regarding the four hundred and four types of identifiable, complex disorders, named in *(6)*: these refer to a standard classification of illnesses in Tibetan medicine.

In *(7)*, a sense of paranoia is described, maybe reactive to this feeling of losing hold: what's happening? When the familiar sense of self is melting, we get anxious and project threat outwards. What happens is shifts in energy, shifts in perceptions, shifts in giving meanings. Nothing is put into the mind, no mantra, no text, no transmission … this is just my, your mind!

In *(13)*: the five poisons are the traditional three, attachment, aversion, and ignorance, complemented with pride and jealousy.

As to *(15)*: there is the mentioning of joys, sorrows, of all kind of experiences that cannot be articulated. *Nyam* may derive from forms of mistaken perception. For example, I remember Wallace explaining that it may be said that a visual appearance of a dragon is valid with respect to that appearance. But, since there is no dragon, that perception is mistaken with respect to its apprehended object, the dragon. Clearly, meditation may arouse all kinds of experiences and emotions, pleasurable and un-pleasurable, from bliss to terror; as I also experienced.

Talking about dragons: in *(27)* we also meet with demons. Wallace recalls how Gen Lamrimpa, in the one-year shamatha training in the US in 1988 (referred to in Chapter 1.2), warned participants that they might experience visions of demons or other terrifying apparitions. Many Tibetan contemplatives have reported this kind of experience when doing intensive and prolonged practice; and the traditional Buddhist worldview includes the existence of these beings. It seemed that no one during the retreat reported seeing any such apparitions that are hardly part of Western worldview. However, some participants were plagued, by – maybe – Western versions, as Wallace illustrates: the demons of lust, spiritual arrogance, fear, boredom, self-doubt, guilt, and low self-esteem. In The Vajra Essence, Düdjom Lingpa poses that all these beings and demons have no existence except as appearances to the mind. Wallace states that this may be the view of many psychologists today. Yet, as he adds, Düdjom Lingpa poses that the self is no more real than these other appearances to the mind. Both kinds of phenomena are "empty" of inherent, objective existence. [10]

Reference is made to the four kinds of mindfulness: the four kinds as mentioned in the Four Applications of Mindfulness sutra, that's very basic to vipassana practice. As we saw, they are: mindfulness of the body, feelings, mental states and reality-patterns. This matrix of vipashyana techniques is usually applied sequentially, after you have practiced shamatha. Here, however, in the course of just doing this Settling the mind shamatha practice, the insights of these four applications of mindfulness may arise spontaneously. Wallace stated that the Settling the mind practice – a practice of shamatha, aiming at bringing us to the substrate consciousness as a foundation for realizing emptiness and *rigpa* – may, as an unexpected bonus give us some very profound insights into impermanence, suffering, and non-self. These are the "Three Characteristics," explored in vipashyana insight practice.

In *(29)* there is the mentioning of a small degree of extrasensory perception: there may be some glimpses of extrasensory perception and higher knowledges (like clairvoyance and clairaudience) that can be achieved when one is much further on the path. An allusion to this has been made in Chapter 5.3. I remember that some of us participants started having dreams with the same content. Two of them, for instance, found out they had dreamt of a specific type of dog in the same night. For another example, one participant heard a word, in the silence of his dream, in a language that he didn't speak. He had been awakened in the middle of sleep by the voice of his neighbor saying this particular word. Recognizing the voice, he asked the woman about it the next morning, spelling out the word phonetically on a napkin. She reported that, while she hadn't spoken, she had been awake and pounding on the wall, fearing that bears had again broken into the dormitory hallway. The word he "heard" was the word for tapping or pounding in her language.

Anxiety

It strikes me as interesting that quite some direct references are made in this section to anxiety, fear, anguish and terror, like in *(21)*, *(22)*, *(24)* and *(27)*. More anxiety seems to be included in more hidden ways, for instance about things that may be harmful, and fear underneath anger. Anxiety seems to be the most central in the experiences described.

Most people pass through many kinds of experiences habitually

– some less, some more turbulent – through the days, through their lives. This includes grasping on to everything that comes up, and in this way perpetuating samsara. Now, with this practice we are allowed to catalyze that which lies between our habitual conscious awareness and the substrate consciousness, and "work through" it, practicing non-grasping; which indeed is a therapeutic endeavor. In meditation we are instructed: don't "believe" the thought, don't go with the delusion, don't grasp to this form of grasping. As it is said, if you attend to the delusion, then it will be reinforced. And the anxiety will increase.

In *The Vajra Essence*, three types of disturbances are named that very likely will show up in the course of shamatha practice for a meditator who wants to take shamatha to its culmination. Wallace describes:

First, there are "outer disturbances such as magical displays of gods and demons." As demons and dragons are not a common part of our conditioning and meaning making in the West, we might rather use images for things without material existence like electromagnetic fields, photons, quarks, and superstrings. As Alan notes, most Westerners are more likely to bump into a superstring than they are to see a demon. What we would find as outer disturbances, on retreat, may be the sounds of neighbors, parking lots, traffic roads, or insects.

Second, there are "inner disturbances including various physical illnesses," coming from outside or inside. There may be viruses at play, in respiratory ailments, but also the elements in our body can fall out of harmony, leading to a rash and itching. I described the rash and itching that I experienced (33); this may fall in the present category.

Third, there are "secret disturbances of unpredictable experiences of joy and sorrow." They may just come up, surprising and even shocking you. You may feel elated, or suddenly very sad, or moved to tears. There's no reason as far as you can see. Secret refers here to: not outer, not inner, but mysterious, without apparent reason or cause.

In general: be sensible, see the teacher, and if really necessary a doctor or a therapist, as mentioned. It is important that we are not fixated on external causes. The risk is that the fundamental issue that this is all arising as a manifestation of our own mind gets out of sight. If we release the fixations, the mind has an immense capacity to heal itself. The path is: continuing without grasping and let the mind settle.[11]

Other voices about signs of progress

While some other authors have also addressed these "signs of progress," this has usually not been in such detail as Düdjom Lingpa. Padmasambhava, in relation with shamatha, gives a description combined with advice. Joining with the general instruction to observe whatever thoughts come up, without grasping, he notes: " … if detrimental habitual thoughts suddenly pop up, focus right on that compulsive ideation as your meditative support. Whatever detrimental, habitual propensities of attachment and hatred arise, recognize them; and in a relaxed way release the mind right upon them." And: " ... if depression or sadness arises, your consciousness has become distorted, so meditate on the disadvantages of samsara, the difficulty of obtaining a human life of leisure and endowment, and impermanence, and cultivate reverence and devotion for your spiritual mentor or your lama." So, he offers the Four Thoughts that turn the mind as an anti-depressive prescription.[12]

Thrangu Rinpoche describes meditators who cannot stop talking, or who cannot hold still and sit down and meditate, who have to get up and go somewhere immediately. He also addresses emotional turmoil including spirits, ghosts and animals. They are not things that come about because of meditation. These things happen because that is the nature of samsara. Both experiences of bliss, luminosity and nonconceptuality and mental turmoil are temporary experiences involving fixation upon the conception of self. The conception of a self leads to the attachment, and tends to get in the way of one's meditation developing further. As he states: "So the absence of that attachment is necessary, and the way to bring about the absence of that attachment is to rest in the mind's way of being."[13]

Even if not many teachers use terms like "signs of progress," it feels like there generally is a relaxed atmosphere around dealing with turbulent *nyam* manifestations. It may not be so much about specific content, but there is a logic to the process, to the possibility and almost inevitability that *nyam* arise. Crucial is non-grasping, non-identification: with these, the arising *nyam* can manifest as a sign of progress.

Here is a modern-day confirmatory voice to these signs of progress as described by Düdjom Lingpa, by meditation teacher Andrew Holecek. He starts with inviting you to celebrate your courage, in beginning to really look! And then: "Be aware that emotional upheaval and so-called psychological problems can be a great blessing; they can be signs of progress and not regress. Celebrate the fact that things could be coming apart as your ego comes apart. Even though from a conventional point of view that's the last thing the ego wants to hear, from a spiritual point of view this is something to rejoice in." [14] The good thing is that we are getting familiar and intimate with who we really are. We are familiarizing ourselves, and habituating with the nature of the psyche and the mind. From coarse to subtle to very subtle, as the direction ...

Expanding to an increasingly spacious perspective: Düdjom Lingpa addresses what may come after this purifying ordeal in Settling the mind practice. Next to the already known elephant, monkey and rabbit, he now introduces a cripple and a blind stallion. He compares the mind with a cripple, and vital energy with a blind, wild stallion. They are subdued by fastening them with the rope of meditative experience and firmly maintained attention: the reins of mindfulness and introspection. Meditators, with disciplined practice, may keep hope: as a result of their experience and meditation, a moment may arise when they have the sense that all subtle and coarse thoughts have vanished. Then, they experience a state of unstructured consciousness devoid of anything as an object on which to meditate. Then, with the guidance of their teacher, their awareness may reach the state of great non-meditation. It is in this context that Düdjom Lingpa remarks: "Thoughts merge with their objects, they disappear as they become nondual with those objects, and they dissolve ... the mind is transformed into wisdom, the power of awareness is transformed, and stability is achieved there."[15] Here, Düdjom Lingpa describes the transition of shamatha – Settling the mind practice, including turbulent phases in the path, to Essential insight practice. We will be back at that in Chapter 7.

6.3 Diaries, and the texts by Lerab Lingpa and Düdjom Lingpa

The instructions given by Lerab Lingpa on Settling the mind in its natural state, and certainly the meditative experiences mentioned by Düdjom Lingpa, will address meditators who, as monks and nuns, are in the process of doing this practice for many hours a day. Possibly they will practice many months, even years in succession. In that context, our three months' training has been very brief, relatively, "peanuts." Consequently, many of the phenomena described will not have emerged in their possible strength. Still, most of us participants never before had the opportunity to practice such a relatively long period in this kind of structured meditation environment. Many of us may have taken one week retreats (as has also been a criterion for participation in the Shamatha Project), ten days, maybe a month retreat, but probably not much more. So, chances are big that all of us at some moment have entered real new un-known, un-imagined territory. As for myself, next to quite a number of week-retreats, around three weeks had been the maximum duration, before my Shamatha Project participation. Indeed, during the second month of expedition, in (30), I wrote down "Feeling: here I'm really in unknown territory, never been here."

Next to the newness, what we've come up with has been quite different for everyone of us. This is also stressed by the two Dzogchen masters. Lerab Lingpa points to the fact that " ... there are great differences in the dispositions and faculties from one individual to another" (6.1). And Düdjom Lingpa makes the urgent advice: "Spiritual friends who teach this path properly must know and realize that those experiences are not the same for everyone, so bear this in mind!" (6.2 nr *28*). A simple distinction is often made, according to one's idea about how meditation is going: either we think our meditation is going fine or we think it is going poorly (and something in-between). In concordance with that, some speak of "good" *nyam* and "bad" ones. The advice is: when having a good one, we should not take too much delight in it, and when having a bad one, we should not feel too bad about it. Both "good" ones, like temporary experiences in the sphere of bliss, luminosity, and nonconceptuality, and "bad" ones may be seen as "signs of progress"! While meditation and realization get deeper, these temporary experiences will gradually be less.

Reading the texts, I've been so fascinated and intrigued by the astounding similarities and resonances in their descriptions, made a few centuries ago (and building on many centuries before that), and the descriptions I read in my diaries. These diary notes have been written before I studied Lerab Lingpa and Düdjom Lingpa's texts. From what I've heard from other participants and from what I see, re-reading my diaries, an important common theme, showing in many forms and guises, felt or suppressed, seems to be anxiety, fear, terror, as noted before. The backdrop of them all, in this context, may have been: anxiety of losing our mind, in the sense of our habitual, familiar mind, our habitual, familiar sense of self. In what follows, I put next to each other, on the one hand, some descriptions of meditation experiences as presented by Lerab Lingpa and Düdjom Lingpa, and on the other hand, by me. While the brief text by Lerab Lingpa is referred to in a general sense, numbers in *( )*

italics parentheses refer to the sections in the text by Düdjom Lingpa. Numbers in non-italics parentheses ( ) are pointing to fragments from my diaries and other experiential data, as presented in Chapter 5.

6.4 Lerab Lingpa: nothing can harm the mind

Alan invited us to ponder questions about the mind, including: can the mind be damaged? (18). It may feel like that, at some moments. As Düdjom Lingpa mentions in his text: the sensation can be that "all your thoughts are wreaking havoc in your body, speech and mind …" *(1)*. Lerab Lingpa's writing about a "nonconceptual sense that nothing can harm the mind, regardless of whether or not thoughts have ceased" has felt to me as a reassuring truth. "Whatever kinds of mental imagery occur – be they gentle or violent, subtle or coarse, of long or short duration, strong or weak, good or bad – observe their nature …" The space of the mind, including its content: this subspace of awareness cannot be harmed or damaged.

Some points in the diaries do connect very much with remarks made by Lerab Lingpa. For me, the sense of the meditation process has much been one of familiarizing. I gave some words to this: more trust, feeling more safe internally. Feels like a "breaking the barriers" of anxiety, inwardly. Little steps, step by step, deconditioning by deconditioning, molecule by molecule. Feeling more at home, familiarizing (19).

Molecule by molecule: we might also say: piece of dust by piece of dust! Breaking the barriers of anxiety: this feels like a shift in the quality of trust from a more relative trust, more depending on circumstances, to getting in touch with a deeper absolute trust, a non-referential trust that's just there, that's our natural condition. It is said, indeed, that: like speaking about Little Mind and Big Mind, the distinction can be made between Little Trust and Big Trust. While Little Mind refers to our everyday-mind, that's involved with organizing and coping, with hope and fear and self-preoccupation, Big Mind acknowledges and celebrates its capacity to awaken. There is confidence in our inherent joy, compassion and wisdom. In a comparable way, one may speak of Little Trust, trust with an object, trust in a person or deity, while Big Trust has no object. Big Trust connects with our inner resources, our capacity for awakening.

"Acting as your own mentor," another expression by Lerab Lingpa, has had an inviting sound to it – maybe sometimes too inviting, for an adventurous girl. Anyhow, I think I profited from it, as shown in (2), where I just explored different options of using "training wheels" in meditation, to help me get going. And certainly this was the case after I fell on my head on frosty Rocky Mountain soil, in the last week of the retreat: I advised myself as a compassionate mentor about how to proceed with the meditations (this I will describe in greater detail in Chapter 9).

A number of times I described a feeling of radiant clarity, so razor-sharp (for instance 43), such precision. Lerab Lingpa, in his continuation of the mentor-line, pin-points this sharp precision: " ... if you can bring the essential points to perfection, as if you were threading a needle, the afflictions on your own mind-stream will subside ..."

6.5 Düdjom Lingpa: intolerable pain

Interestingly, the text by Düdjom Lingpa is presenting experiences in a thoroughly phenomenological way, practically leaving out, in this section, cultural context and Buddhist meanings. While the aspects he names do so much concur with experiences that I described, it seems that sometimes even we were using "the same words" (of course, with in between, a process of translation). I started to make some pairings, then subdivisions came in, the physical, the mental; then sub-subdivisions. Below, I have placed some categories of experience, as described by Düdjom Lingpa, next to mine. For practical reasons, in the sequence below I first name diary fragments as presented in Chapter 5, and then his text. Many more examples could be given! A way of organizing presented itself in five groupings. There is body-energy, then attachment, aversion, and ignorance: the traditional three poisons of suffering. The last group includes clarity, cognizance and joy. I notice that quite some items can be placed under more than one heading; like, both craving and anger will always be based, also, in ignorance. For various items, mention is made under the particular heading that seems most relevant. Headings in representation of the diaries are with augmenting numbers: meaning that they follow the time line from start of the expedition to end. Italics are added for easy recognition.

Body-energy

Distinguishing three aspects: physical pain, body-mind link, and energy dynamics seems to cover most of the phenomena addressed.

Physical pain

There is much mentioning of physical pain, in many ways.

In the diaries for instance I address: (10) muscular shocks and *pains in my legs;* (33) extreme tiredness, a sense of paralysis, *pains throughout the whole body …*

I find quite some mentioning of pain by Düdjom Lingpa in his *Vajra Essence* text: *(2)* a *sharp pain* in your heart, *(5) intolerable pain* throughout your body, *(27) carrying away your head, limbs and vital organs.*

Body-mind link

There is quite some referring to the body-mind link, including in perception, thinking, emotion.

When looking at my diaries, I read: (9) my hands feeling like *claws, grasping, with clamping bent fingers,* and as if having a greater size … Sort of perceptual distortions; (31) A horn blowing, I hear … This sense of threat – is there really something outside? ; (43) *moving to tears.*

In Düdjom Lingpa's text I also read some fragments, where body and mind are clearly connected: *(1) all your thoughts are wreaking havoc* in your body, speech and mind; *(25) Weeping* out of reverence and devotion.

Might there be a relation of, on the one hand, my mentioning in the diary (35) of the lunch car passing, having a clear visual image of what's in it, and getting later confirmation; and Düdjom Lingpa's reference in *(29)* to the phenomenon that one may develop: "a small degree of *extrasensory perception*"? Golden-brown mozzarella crust ... This, in a way is also about body-mind link, and then about its transcendence.

Energy dynamics

In this Body-energy subdivision, there are the energy aspects, including shifts in energy flow, with bodily manifestations:

In the diaries, I find descriptions; and while writing now, I'm getting the felt-sense again! (4) I'm aware that the *range of contraction and relaxation is so wide* ... There's so much with energy in this practice!; (10) sudden *hot flushes* (while not being menopausal); (33) extreme tiredness, *sense of paralysis* ... needing ten hours' sleep.

In this context, Düdjom Lingpa mentions: *(9)* you think your heart will *burst*; *(10) insomnia* at night, or fitful sleep like that of someone who is critically ill; *(14) speech impediments and respiratory ailments.* When Düdjom Lingpa mentions speech impediments, I'm reminded of my suddenly speaking Dutch, as if escaping me (10). Maybe not an impediment (although not very practical), but certainly a *nyam*!

Paranoia as a sign of progress

In this section, I present coinciding phenomenological descriptions of experiences that find their base in attachment, aversion, and ignorance, the three roots of suffering. These are often pictured in Tibetan iconography as, respectively, a rooster, a snake and a pig.

Attachment, greed, compulsiveness

In the diaries, I find quite some variations on themes of attachment, in various degrees of strength: (1) "think out a thought," with some *grasping* ... *a tenacious compulsive habit* ... this thought may be so important, that it "has to be completed" (15) chocolate bars ... with mind-body bringing in the physical. *Desire, inner taste* ... envy! Pang in the belly.

Düdjom Lingpa also has seen many variations with his meditation students: *(12)* the conviction that there is still *some decisive understanding or knowledge you must have*, and yearning for it; *(16)* that must be a sign ... for me ... *compulsively speculating* about the chirping of birds; *(24)* uncontrollable ... *obsessive attachment.*

Aversion, anger, fear, paranoia

These four can easily go together. The theme of anxiety shows up rather thoroughly. Indeed, with these aspects, strikingly, the diaries abound: wide-ranging examples, in colorful details. A selection: (20) so often not being "now": as if always having to check forward, backward ... *underlying: anxiety,* so habitual; (30) Comes in a feeling-flash as if it is done purposefully "against me" ... I'm sought after ... a flash of *paranoia*; (30) *sabotaging, ridiculing* ... (at first to other, then to self).

Düdjom Lingpa gives the following descriptions: *(18)* unbearable *anger due to having paranoid thoughts* that everyone ... *(19) negative reactions* ... retaliating verbally, *(24)* uncontrollable *fear, anger* ... *hatred.*

Ignorance, disorientation

Ignorance is underlying all other categories. A lot of ignorance I am not aware of. The ignorance that I'm aware of may go hand in hand with feeling a need for understanding or knowledge. Conscious and unconscious ignorance and certainly disorientation often won't be far removed from some kind of anxiety. In the diaries I find: (7) I feel like I'm *orienting myself in spaces,* (36) disorientation. Recurrent themes (in dreams): not recognizing, feeling lost, *disorientation,* (50) (in a dream) *a lot we don't know yet* ... some agitation, worry.

Düdjom Lingpa writes about: *(11)* disorientation; *(12)* some decisive understanding or knowledge you must have.

Diary after retreat, 2010

> Doing the Settling the mind practice we had a good chance to experience distractibility, discontinuity, and built-in reactivity, time and again. The reactivity referring to the movement toward – and away from –, connecting with the three roots of suffering, attachment, aversion and (underlying) ignorance. The three roots: standing for a "bias in information processing" that we carry with us from a very young age!

Clarity, cognizance, joy

For a rounding up of the overview: descriptions of these named experiences are present as well.

In the diaries: (22) this underlying experience of gentle joy, all the time. And: (41) glimpses in *bliss, clarity-luminosity, cognizance,* glimpses in *knowing things to their nature*; (43) *this "being knowing,"* timeless, still. A joyful silent *beaming kind of excitement, gratitude.*

Düdjom Lingpa describes these experiences in his students: *(26)* the saturation of your mind with *radiant clarity and ecstasy,* like pristine space; *(27)* you experience a sense of *ecstasy* as if the sky had become free of clouds; *(29)* an extraordinary sense of *bliss, luminosity and nonconceptuality.*

These are striking parallels. What sounds parallel are wordings, phenomenological descriptions referring to experiences, mediated by words. These in some sense and to some degree turn out to be universal, through time and space! I find this amazing. Passing through translations, through my enthusiastic associations – still, parallels, in some sense.

6.6 Contextualizing experiences, then and now

How about the cultural perspectives of authors and translators? A little pointer to possible differences, when we not only look at direct phenomenological experience but also at wordings of how a person is giving contextual meaning to these, is showing when we take some of the brief comparisons presented in a lengthier form. Certainly, the brief version of a description of bodily pain, as in my "extreme tiredness, a sense of paralysis, pains throughout the whole body" (Chapter 5.3, nr 33) might also be described by Düdjom Lingpa's meditation student. And the student's, as described in Chapter 6.2, *(27)* "carrying away your head, limbs and vital organs" could be recognized by me.

However, when we take the full description, a different picture comes up. My description is embedded in a comparison with "cold turkey," as I imagine that to be; like waning from addictions, neuro-transmitter shifts, body-mind feelings of withdrawal of chemicals. What the student describes includes the feeling that gods or demons are actually carrying away his head, limbs and vital organs, leaving behind only a vapor trail. So, with comparable phenomenological descriptions (Experiences perspective) here we see quite different cultural interpretations that belong to the time that the text evolved (Cultures perspective). Both expressions, by themselves, are equally valid. It is just that our Western interpretative framework has mainly left gods and demons behind. Cold turkey will not be a metaphor prevalent in nineteenth-century Tibet.

Another aspect I notice: when I imagine Düdjom Lingpa's students in the nineteenth century, and when remembering retreats that I took part in, in the twenty-first century, there is a difference in what I see. As if automatically, with Düdjom Lingpa I see before my eyes a group of men. This may probably also be inspired by old photographs. In contrast, in the present, I see a mixed women-men group. Yes, the situation for a woman yearning for Dharma practice must have been quite different, in nineteenth-century Tibet, as compared to now, here. Let me take a brief look to the four Integral Perspectives on meditation, in this gender-context. As for the Experiences perspective, the Tibetan woman meditator and I would have a lot in common, in practicing shamatha. Our direct phenome-nological experiences may be quite similar: with tiredness, pain, itching, fear, peace, bliss. However, our way of contextualizing these experiences will be quite different. As to Cultures, she and I live in very different worlds. In her context, traditional full monastic training is closed to women, even while it is maintained that they have equal access to enlightenment. As to Behaviors, while I was sitting in a field laboratory being hooked up to a computer, she may be sitting outside in the mountains, where no digitalized world yet evolved. And regarding Systems: for instance, as a housewife, she would live within traditional androcentric, patriarchal structures, with probably not much opportunity for formal practice. Of course, there are a few exceptional women that are widely known and cherished as highly evolved teachers: like, for instance Yeshe Tsogyal. She, fully enlightened, was Padmasambhava's consort, living in the eighth century. There was Sukhasiddhi (eleventh century), who lived a life of poverty as a mother with six children, who became a generous business woman at age 59, then received empowerments and instructions, and attained full enlightenment at age 61. More recently, there was Ayu Khandro, living in the nineteenth and twentieth centuries, having reached an exceptionally high level of realization. They are exceptions. Such has been the situation.

How about the "natural state"? Our experiences, perceptions, interpretations: all, in this Tibetan Buddhist view, are appearances arising from the substrate consciousness. So, this relative natural state, in a student from Düdjom Lingpa, or a woman meditating on her own, will look different from mine. The ground of the relative psyche, when more superficial personal meanings, labels, interpretations, stuff, defenses, neurotic dynamics are stripped off (and maybe many of us have a lot more of that than Düdjom Lingpa's student), will still be different. This is the case because the mental continuum, in this view, is an individual stream of consciousness that carries on from one lifetime to the next. So, also in this way we come to the understanding: the relative natural state in me, now, and the student, then, and others, women, men, now and through the ages will vary. At the same time the ultimate, absolute natural state is one for all of us, always already; how could it be otherwise ...

Which brings us to Chapter 7, addressing Settling the mind practice and the natural state in the context of Mahamudra and Dzogchen.

Chapter 7

Settling the mind, Mahamudra and Dzogchen

In Natural Liberation, an important treatise ascribed to Padmasambhava, we find the following text about the practice: " … settle the mind in its natural state by letting it be just as it is, steadily, clearly, and lucidly … in the mind's own mode of existence." (7.2)

Of the four main practices that Alan Wallace taught during the expedition – with the focus on breath, mind, awareness, and heart, respectively – most of my time has been devoted to breath practice with focus in the abdomen, and to Settling the mind practice. While shamatha practice overall leads to observing and focusing in progressively refined ways, Settling the mind directs the attention, more and more subtly, to mind and mind processes.

As Alan Wallace repeatedly stated, while the Settling the mind practice certainly yields many insights into the nature of the mind, it is rightfully classified as a shamatha-calm, and not as vipashyana-insight practice: this practice doesn't include any investigation into the nature of phenomena. It can lead to an experiential realization of substrate consciousness, but it does not lead to realization of ultimate truth. Nevertheless, Settling the mind's special advantage of leading to insight into a variety of mental states and the relative nature of the mind makes the practice really exciting and challenging.

Settling the mind in its natural state practice is considered a direct preparation for the Essence traditions Mahamudra and Dzogchen. After shamatha adventures, and having participated in a number of Mahamudra and Dzogchen retreats in recent years, now that I have a more contextual felt sense, my love for Settling the mind practice has only increased. How does that relate to my earlier experiences, including as laid down in the diaries, in Chapter 5?

In this chapter we dive more deeply, specifically, into Settling the mind in its natural state as it has been described and contextualized in the Essence traditions. While this goes for every chapter, here even more I'm aware that this is about my humble limited understandings, at the moment of writing.

7.1 Some names, notions and context for the practice

While the practice Settling the mind in its natural state (Tibetan *sems rnal du babs*) is drawn from the Dzogchen, or Great Perfection lineage, according to Wallace, it is found in other Buddhist traditions as well. The practice has been referred to by various names in English translation, including: "meditation on the relative nature of the mind," and "meditation on the conventional nature of the mind," as The First Panchen Lama (seventeenth century) called it. Düdjom Lingpa called the practice: "taking ... the mind as one's path" and "taking appearances and awareness as the path." I like this last one as, in describing the path it includes both the space of the mind and what arises in it. Karma Chagmé, seventeenth-century teacher, himself known for his important synthesis of Dzogchen and Mahamudra, quotes Maitripa, who lived in the eleventh century: Maitripa spoke of a comparable practice as "quiescence focused on conceptualization," and "transforming ideation into the path." [1] All these various namings by various masters make clear that the practice has been around for many centuries. As a reminder: Maitripa described three types of shamatha meditation in the Essential context, as we saw in Chapter 4.1: first, quiescence that depends on signs, second, quiescence focused on conceptualization, and third, quiescence that is settled in nonconceptualization. So, indeed, here we take conceptualization as the focus.

Dzogchen and Mahamudra, for which Settling the mind is seen as the shamatha part, in their own ways both focus on the realization of the nature of consciousness. Dzogchen (The Great Perfection, or The Great Completion) has been mostly associated with the Nyingma lineage in Tibetan Buddhism, and Mahamudra (The Great Seal, The Great Symbol, The Great Embrace) was mostly connected with the Kagyu lineage. While there are differences in terms of their methods, in terms of realizing the true nature of phenomena both the Dzogchen and Mahamudra teachings are said to be the highest practice, Thrangu Rinpoche notes. [2] While there has been a longstanding Dzogchen Bön tradition, these practices have also found their ways into other schools: to Nyingma and Kagyu and also to the Sakya and Gelug schools and into the non-sectarian Rimé movement.

Alan Wallace, previously ordained in the Gelug tradition, told us that he was taught Mahamudra – with a shamatha and vipashyana part – by Geshe Rabten, a Gelug teacher. The technique of Settling the mind in its natural state that Wallace teaches is identical to the shamatha part of Mahamudra that Geshe Rabten taught him. Since then, Wallace received similar teachings from the late Tibetan contemplative Khenpo Jigmé Phuntsok, and from Gyatrul Rinpoche, both from the Nyingma tradition. Another teacher that Wallace often refers to in his writings on Settling the mind practice is Gen Lamrimpa, from the Gelug tradition. [3]

Have a sense of the Settling the mind practice again in a guided meditation, and feel for yourself which naming best attunes with your experience.

Settling the mind, releasing grasping – guided meditation

While we did this practice before, with instructions as described in Chapters 2.3 and 5.3, following here is Alan Wallace's guidance from a later date during the Shamatha retreat in September 2007, with some different accents. Let's build the practice up, the felt-sense of a "mind, settling" being introduced with settling the body in its natural state, and the breath in its natural rhythm.

> Make it a gentle entry as you bring awareness to the body, softly permeating the whole field of sensations and immediately

feeling a release, a release of tension that may have built up, wherever it may be, in the shoulders, the face, the jaws and especially the eyes. As you settle the body in its natural state, the breathing in its natural rhythm, rest your mindfulness in this field of tactile sensations. Attend to these tactile events without distraction and to the best of your ability without grasping.

We come to Settling the mind in its natural state. It may be quite a roller coaster if you spend days and weeks doing it. It is quite an adventure.

Let your eyes be at least partially open and rest your gaze vacantly in the space in front of you. The space need not be limited by the proximity of visual images in front anymore than a rainbow obscures the sky. The visual appearances do not obscure the space in front. You may rest your gaze in front of you, in space. Rest it vacantly and now turn the full force of your mindfulness to that domain of experience, the sixth sense domain, which is the domain of uniquely mental events, for discursive thought and mental images, desires, memories, emotions, the whole range of purely mental events which you detect only with mental perception and not with any of the five physical senses. What's left over when all of your five physical senses have gone dormant? It's the domain in which dreams arise and into which they pass. If you are new to this practice, you may find that domain elusive, in which case, as a great preliminary exercise, you may deliberately, rather ponderously generate a discursive thought. It could be any thought: there is a mountain, this is a teacup, or this is the mind. We are not attending to the referent of the thought. Focus your full mindfulness upon the thought itself, as syllable by syllable you bring it to mind. Allow the thought to vanish back into the space of the mind and keep your attention riveted right where it was, attending closely. And witness the next thought, mental image or other mental event that arises spontaneously. And just let it be, attend to its nature *without distraction, without grasping*. Observe it just as it is, without intervention.

If you lose your orientation, become a bit spaced out, or simply get caught up in thought, you may once again deliberately generate a mental event, this time perhaps a mental image of anything you like: bring it to mind. Focus single-pointedly on it until it fades out and keep your attention right where it was, placed in this space of the mind, observing whatever arises within that space.

Without distraction means: without getting caught up in the referent of the thoughts, carried away from the here and now. Without grasping: without the superposition of "I" or "mine," without the superimposition of "I like" and "I don't like," without even preference that the contents of the mind may subside. Rest in the larger space of your awareness as you attend to the subspace of the mind. Rest in stillness, in the space of your awareness even while your mind is active. Unlike other shamatha practices, here we do not banish or let go of the thought, we do not even prefer for there to be fewer thoughts rather than more. Simply release the grasping onto the thought and observe whatever thoughts arise in the mind with unwavering mindfulness. Now and again apply introspection.

Let's bring the session to a close.

"Rest in the larger space of your awareness as you attend to the subspace of the mind," with this Wallace invites you to rest in the space of awareness that transcends the individual subspace of the mind. This is about the space of awareness that the practice Awareness of awareness increasingly connects with.

7.2 Settling the mind: mind and awareness

When I started reading more about settling the mind in texts of the various traditions, I wondered and felt some confusion: it seemed that the notion "mind" was used in so many ways. Was it referring to ordinary everyday mind, to mind space, to purified mind, to awareness in general? In Tibetan language there are many more fine differentiations in terms and meanings for mind and awareness than exist in Western languages, we in the West lack the refined vocabulary. Next to that, the meaning of many notions and technical terms may be quite different across the many traditions of Sutra, Tantra and Essential traditions. Different contents and meanings of notions regarding mind, thought, conceptuality, awareness, space, and different namings in the Dzogchen and Mahamudra context, do contribute to confusion in translated terminology.

I learnt that one important distinction, in Dzogchen, is between mind (Tibetan *sem* or *sems*, Sanskrit *citta*), and awareness (Tibetan *rigpa*, Sanskrit *vidya*). Sogyal Rinpoche makes this distinction: *sem* refers to the "discursive, dualistic, thinking mind, which can only function in relation to a projected and falsely perceived external reference point." As he clarifies: *sem* is "unstable, grasping, and endlessly minding others' business; its energy consumed by projecting outwards ... It is within the experience of this chaotic, confused, undisciplined, and repetitive *sem*, this ordinary mind that, again and again, we undergo change and death." *Rigpa* is described by him as "a primordial, pure, pristine awareness that is at once intelligent, cognizant, radiant, and always awake." This is about "the very nature of mind, its innermost essence, absolutely and always untouched by change or death." I feel moved by Sogyal Rinpoche's poetic description, about this innermost essence: "At present it is hidden within our own mind, our *sem*, enveloped and obscured by the mental scurry of our thoughts and emotions. Just as clouds can be shifted by a strong gust of wind to reveal the shining sun and wide-open sky, so, under certain special circumstances, some inspiration may uncover for us glimpses of this nature of mind." These glimpses know many degrees of depth, as he says; however, each of them will bring some light of understanding, meaning, and freedom, because the nature of mind is the very root itself of understanding. [4]

Thrangu Rinpoche, from a Mahamudra perspective, clarifies terminology in this way: "The instruction from the Dzogchen tradition reveals the essential point that makes this instruction section of ... Dzogchen teachings unique: distinguishing between mind and awareness. In the context of that tradition, the term mind is used to refer to what we would normally call thought or deluded mind." With "we" he points to: we in the Mahamudra tradition, where mind in the sense of *sem* is referred to as thought or deluded mind. About Dzogchen, he continues: "Awareness is used to refer to the innate nonconceptual cognitive lucidity of the mind. The point made in this tradition is that it is of great importance in meditation to properly distinguish between these two in your meditation experience." As Thrangu Rinpoche explains: for Mahamudra, when it is said that we "meditate on the mind," we are referring to the true nature of mind, or the way the mind is. "The reason for teaching meditation on the true nature of the mind is that all phenomena are just mind."[5] So, in this Mahamudra context, "mind" in a wider sense refers to primordial awareness.

Settling the mind in its natural state: the terminology that Wallace uses generally includes mind (*sem*) and awareness (*rigpa*) in Dzogchen sense. In *Natural Liberation*, an important treatise ascribed to Padmasambhava, we find the following text about the practice: " ... settle the mind in its natural state by letting it be just as it is, steadily, clearly, and lucidly ... in the mind's own mode of existence."[6] What mind is addressed here? The quote comes from text in the shamatha section. In this Dzogchen text, mind appears to refer to natural relative mind, in quiescence context. Yet, absolute mind is always shining through relative mind. In what follows, the context will make clear what is the intended meaning for mind.

7.3 Shamatha, awareness and vantage point

In my explorations regarding how to view shamatha, and specifically Settling the mind shamatha practice in the context of the Essence traditions, some basic descriptions and elaborations by Mahamudra teacher Daniel Brown have been very clarifying for me. As he explains in *Pointing out the Great Way, the stages of meditation in the Mahamudra tradition,* according to the Essence viewpoint, three main possible levels of awareness are distinguished:

The *coarse* level stands for the habitual mental content in the stream of consciousness, like thoughts, sense perceptions, and emotions,

The *subtle* level refers to the fleeting mental activity surrounding sensory experience before that activity becomes full-blown mental content,

The *very subtle* or extraordinary level: this is the level where impressions from actions in the past are accumulated before they ripen in fresh experience. This very subtle level is sometimes referred to as storehouse consciousness. According to Daniel Brown: "The point of observation of this storehouse consciousness transcends our ordinary sense of self and individual consciousness. Like a vast ocean of awareness, this vantage point for the extraordinary meditation is typically referred to as the *always-here mind* or *awareness in and of itself.*" And he continues: "When the mind operates primarily at this very subtle level, the individual is significantly more prepared to realize the mind's primordial nature, which is always there unaffected by all the mental activity at the coarse and subtle levels. Thus, the Essence perspective cuts right to the heart of the mind's natural state and invites the practitioner to awaken to it in direct experience, the result of which is enlightenment."[7]

The heart of the mind's natural state, I love this expression. Storehouse consciousness (Sanskrit *alaya-vijnana,* Tibetan *kun gzhi rnam shes*) is what Wallace has been naming substrate consciousness or the subtle continuum of mental consciousness, in our late afternoon exchanges. Thrangu Rinpoche refers to it as "all-basis-consciousness," as it is the basis or ground for the arising of all other types of consciousness, the ground for the sensory consciousnesses like hearing and seeing, and the mental consciousness. It is constantly present, constantly operating.[8] Next to these levels of awareness, it is important to distinguish two perspectives from which we may do our practice.

Perspectives

In Settling the mind, the initial instruction is to practice observing the mind as still, and observing mental events. So, these are in a way two kinds of focus. Alan Wallace guided us in seeing that we can shift between the two: focusing more on the foreground of events, or more on the background of space of mind (Chapter 2.3). Next to that, continuing on the thread of one's ability to shift: Settling the mind shamatha practice can also be said to support a shift in perspective. It means that, next to focusing on whatever we focus on, we can train to view from the perspective of unfolding mental events in the mental continuum, on the one hand, and from the perspective of mind, on the other hand. Alan Wallace has invited us, in the practice just presented, to rest in the larger space of your awareness as you attend to the subspace of the mind. He invites you to rest in stillness, in the space of your awareness even when your mind is active; this seems to refer to the vantage point of a more subtle awareness, in comparison to being jerked around with foreground events and content.

This relates to two perspectives, from which the effects of meditation can be described. Daniel Brown refers to them as: the *event perspective* and the *mind perspective*. For preparatory contemplation and concentration practice, the aim, in the mind perspective, is that the mind increasingly stays on the intended object without distraction. From the perspective of observable events, the events in the unfolding mental continuum become calm. So, for a little exploration: when in the coarse state of the mind, when we start practicing concentration meditation with our habitual consciousness, what do we observe? What can be observed from the perspective of the mind? Habitual mind cannot stay on its intended object for even a few moments. And when it stays, this is only partially, with the attention divided between the intended object and the background noise of habitual mental activity. Habitual mind is easily distracted. The awareness of the unfolding stream of experience is discontinuous and there are phases of mindlessness or unawareness. From the perspective of the unfolding events, our ordinary confused mind is not organized and calm. Brown points out that a mind that gets distracted from the intended focus and gets lost in the construction of more and more mental contents can be said to become "elaborated." The various kinds of cognitions that are elaborated are: physical sense perceptions, emotional states, and conceptualization, such as thoughts and memories. So, this is about what arises, the contents, the event-perspective. In the words of Daniel Brown: elaboration is to thought content what distraction is to awareness.[9]

Diary after retreat, 2009

> While practicing Settling the mind shamatha, I get to seeing events, that sometimes have continued already for a time, before I'm aware. The possible sequence, with starting from a continuum of thought elaboration, may go like this, in principle: when I become aware of distraction, a daydream may have gone on for quite a stretch. When I have practiced for a longer time and get more focused, with refined introspection-reflexivity, there may still be quite complex thought sequences. With more training and quicker introspection-discovery, I find thoughts, associated in chains. Gradually I get faster in cutting the chain and going back to the intended object of mindfulness. There is detection of specific thoughts. Then, with quicker and quicker discernment, fleeting thought is caught, and then unelaborated mind-moments. I've described some of these thought-chains in my diaries, like (15) and (30). Exhausting, just to read them now, with this initial heaviness. Then, gradually quicker, lighter, into energy patterns.
>
> At the time of retreat I was sitting in the quiet of a context of many weeks' practice, in mountainous area, with rocks, grasses, huge skies, space. Now I'm back home, with meditation hours squeezed within a busy urban-activities life with agendas and clocks. Still, this inner space of mind and events is here, with a

degree of staying and calm. The inner practice can be continued, cherished, further refined, in an adapted discipline.

The effects of initial contemplation and concentration practices, at the coarse level, in the event-perspective, are that cognitive elaborations will greatly diminish in frequency and complexity. There is more order and calm in the stream of events. From the mind perspective, awareness is increasingly staying on the intended meditation object. After intensive practice, when awareness of each discrete moment of arising is uninterrupted, this is called one-pointedness. This awareness of the spontaneous unfolding of the mind free of cognitive elaboration is said to be a step toward understanding the real nature of the mind. As Daniel Brown notes, in this context: "From the mind-perspective, what emerges is awareness-itself, no longer obscured by conceptual elaboration. Such primordial awareness is unimpeded."[10]

Awareness-itself, reflexive awareness

Reference has been made to "awareness-itself." After the clarification of the use of the terms mind and awareness, I like to briefly address this notion "awareness-itself," in this context of the Essence traditions. This awareness is also referred to as "reflexive awareness," and self-cognizing awareness (Tibetan *rang rig*, Sanskrit *svasamvedana)*. This awareness, according to Wallace, is conceived of as an infallible, nonconceptual, nondual perception of mental phenomena. It is said to be of the same nature as the mental events that it apprehends.[11] Daniel Brown describes awareness-itself as an intrinsic property of the mind that is reflexively aware of its own nature all the time. As Brown states, ordinary special insight meditation on emptiness helps a practitioner determine that there is no substantial, self-existing, inherent self that can serve as the point of observation during meditation. In Mahamudra, it is awareness-itself, instead of ordinary self-representation, that serves as the vantage point of observation during meditation.[12]

When practicing shamatha – even if we do a little insight practice on mind, as Alan introduced us to – we have not yet had the opportunity of determining if there is a substantial inherent self. For that, more prolonged insight practice is needed. Still, even in the earliest forms of shamatha-introspection or meta-awareness a type of *reflexivity* can be recognized, as it is meant to involve an awareness of the state itself. Wallace often referred to this little "corner of insight-vipashyana practice" within shamatha. And of course, in Settling the mind practice, with getting some insights in one's psyche, a certain kind of self-reflection is at work, and cultivated.

The notion of reflexive awareness is addressed in an interesting way, I find, by Antoine Lutz, John Dunne and Richard Davidson, researchers and Buddhist scholars that I will refer to more extensively in Chapter 10. They describe this fundamental form of reflexivity as an explicit aspect of all cognitions. However, most of the time this reflexivity is obscured. Practice with intentionally applying mindfulness, like we do in shamatha with limited monitoring-reflexivity, can help dissolve some obscurations, but cannot recognize the nature of mind. In the voice of Thrangu Rinpoche: "Mindfulness, as useful as it is, contains some conceptual dregs or impurities. The actual recognition of the nature of mind is distinct from mindfulness. It is a cognition without object, a cognition that is recognizing or experiencing itself."[13] While mindfulness always has an object, (reflexive) awareness does not have an object. It is beyond subject and object. As Lutz and his group suggest, in vipashyana, reflexive awareness is increasingly cultivated, and in the Essence traditions this aspect is emphasized to its furthest possible point. They pose the view that the Essence tradition practices differ from other meditations in that, theoretically, in these traditions an implicit aspect of all cognitions – this fundamental form of reflexivity – is taken and made phenomenologically accessible to the practitioner.[14] With these conceptual clarifications, let us return now to first person practice.

7.4 Shamatha practice in Essential context

Many teachers emphasize that the importance of shamatha and a good degree of stability for the Essential practices cannot be overestimated. These are Padmasambhava's words, about shamatha and pristine awareness: "Without genuine quiescence arising in one's mind stream, even if awareness is pointed out, it becomes nothing more than an object of intellectual understanding; one is left simply giving lip-service to the view, and there is the danger that one may succumb to dogmatism. Thus the root of all meditative states depends upon this. For this reason, do not be introduced to awareness too soon, but practice until there occurs a fine experience of stability."[15] So take care that you practice a good deal of shamatha before entering what are called the "higher" practices.

Regarding how far this stability needs to go, for being able to practice in the Essence traditions, there are various views. Traleg Rinpoche, contemporary teacher who died in 2012, states that according to the Mahamudra tradition, it is not necessary "to go through the different levels of concentration and absorption in shamatha meditations. Instead, it is sufficient that we stabilize the mind." We don't need the highest states of concentration or obtain any level of absorption. [16] Lutz and his group note that, in the Essential traditions practitioners of shamatha usually develop a lesser (and often unspecified) state of concentration before being instructed by their teachers to move on to other practices, which no longer specifically involve focusing on an object. What *is* emphasized, throughout the process in consecutive practices in the Essential practices, is the importance of close guidance by a teacher.[17]

Diary after retreat, 2011

> It seems, as I understand it, that there are different views in contextualizing shamatha in the various meditation paths. How far to "let the mind settle itself," by whatever shamatha practice, in this relative natural state? There is the possibility to aim at

> achieving full shamatha, as an intermediary aim, after which various paths are open. One may continue concentration-calm practice into the absorptions, or practice further on the bodhisattva path, or take full shamatha as a basis for insight practices for enlightenment. Also, one can learn from Settling the mind practice about relative mind. However, as I understand, in the Essence context of the deep urgency for ending suffering among beings and the flourishing of positive qualities, quite some teachers don't seem to see the need to attaining the highest stages of shamatha on the Elephant Path. Go as quickly as possible for full awakening, with only the necessary stability practice in this context! This, I learn, is the view of all the Vajrayana traditions in general, including all lineages.

At this point, I would like to share a larger map by Daniel Brown, presented in his book on the stages of meditation in the Mahamudra tradition, as it has been so supportive for my understanding. Just some terms and notions are given here, for possible recognition, and not to be explained more extensively. This highly summarizing map connects the named levels of awareness with the various practices, referred to earlier in this chapter. As Brown states, mastery of the concentration (shamatha) stages of meditation opens up the subtle level of the mind. Mastery of the special-insight (vipashyana) stages opens up the very subtle or extraordinary level of the mind.[18] Then the practitioner begins a new set of exercises, called the extraordinary practices. They build upon the foundation of the ordinary concentration-calm and special-insight practices. These extraordinary practices are especially designed to reveal the real nature of the mind. Two stages are described in the Mahamudra of extraordinary meditation: the yoga of one taste, and the yoga of nonmeditation. Both are designed to set up the conditions to generate enlightenment. Extraordinary practice stands for a coupling together of relative truth and absolute truth, in which both are held in a nondual relationship.[19]

In Dzogchen context, shamatha and vipashyana are principal practices that are necessary before true Dzogchen starts: practice becomes truly Dzogchen only when it reaches the level of nondual practice, as Dzogchen teacher and scholar Namkhai Norbu explains. The same can be said for Mahamudra. In Dzogchen, one then continues, according with two levels of practice. These are: *Tregchöd,* literally: "(spontaneous) cutting of tension," and *Thödgal,* "surpassing the utmost."[20]

Well, however precious and helpful, a map is a map. With the advice: rather practice shamatha, and on, and see for yourself, than looking at a map for too long a time!

7.5 Mind practice, on the way: transcending and including

The former sections about levels of awareness and perspectives have mostly given context to concentration practices in general. In the present section it is especially mind practices, including Settling the mind in its natural state, that are addressed. In the descriptions that follow we may get a glimpse of the "observing" in precious shamatha mind-practices "meeting with," or even "shifting into" vipashyana insight, so to say ...

According to Thrangu Rinpoche, in the Kagyu tradition, the basic format of presentation of the mind's nature is called "abiding, moving, and awareness." He explains that in this tradition, the situations of abiding (the mind that rests) and moving (the presence of thought within the mind) are used to enable students, through their own exploration, to recognize the mind's nature. Students are taught first to look at the mind when it is still, and see how it is then; next they are taught to look at the mind when thought occurs, and see how it is then. With this approach, that includes *both stabilizing* the mind (concentration-calm) and *attaining insight* into what arises, Thrangu Rinpoche states, some students will recognize the mind's nature and some will not; but this is still the most stable, the best way to proceed.[21]

So, this may start with observing the space of the mind and what arises in it, but it is also about recognizing: the combination leads to a more stable recognition of the mind's nature, Thrangu Rinpoche suggests.

Diary after retreat, 2008

> Reading Thrangu Rinpoche's instructions to his students regarding mind practice, about first observing the background of stillness, then what arises: how striking that, in this Mahamudra-vipashyana approach, he advises students to "first observe the mind when it is still"! In the shamatha practice with Alan Wallace, we were often first invited to observe what arises, as this, in the beginning, seems to leave hardly any space for stillness ... let alone, "awareness." So, observing the mind when it is still can only be practiced when there is sufficient shamatha stability, which takes time, for extra-overstimulated-overloaded Western minds ...

Thrangu Rinpoche states about Mahamudra shamatha and vipashyana meditation, in this mind context: "The difference between shamatha and vipashyana is not great; it is only a slight shift." With shamatha practice, our mind comes more and more to rest naturally in a state characterized by less thought and conceptuality. We are able to reduce disturbing emotions through shamatha, but not to eradicate them. Then, Thrangu Rinpoche notes, with vipashyana, the quality of luminosity is enhanced, and we realize the mind's lack of any inherent nature. Through the realization of this state of luminosity and emptiness, we can abandon delusion at the root. [22]

Mind, shamatha and vipashyana

Next to Thrangu Rinpoche in Mahamudra context, also Düdjom Lingpa, in Dzogchen context, sketches a wider perspective, with a shift from concentration to insight. His basic Settling the mind in its natural state shamatha instructions have been presented in Chapter 6.2. In the initial shamatha section in *The Vajra Essence* he describes deeper possibilities of this practice in an experiential way:

> … someone with enthusiastic perseverance may recognize that this is not the real path, and by continuing to meditate, all such experiences of blankness, vacuity, and clarity tainted by clinging vanish into absolute space, as if one were waking up. After this, outer appearances are not impeded, and the rope of inner mindfulness and firmly maintained attention is cut. Then one is not bound by the constraints of good meditation, nor does one fall back to an ordinary state through pernicious ignorance. Rather, ever-present, translucent, luminous consciousness … shines through. Without dichotomizing *self* and *object*, such that one can say "this is consciousness" and "this is the object of consciousness," the primordial, self-originating mind that has experiences is freed from clinging. [23]

Alan Wallace, in his commentary to this text, clarifies that it is here that we transcend shamatha and the substrate consciousness, breaking through to pristine awareness. Düdjom Lingpa continues: "Thoughts merge with their objects, disappearing as they become nondual with those objects, and they dissolve … the mind is transformed into wisdom, the power of awareness is transformed, and stability is achieved there."

After stability in a relative sense, won with effort in a dual context with subject and object, this now may be referred to as an absolute effortless stability. Interestingly, Düdjom Lingpa refers to the named shift in awareness "as if you were waking up," which seems to point to a wider realization, to a shift in the range of coarse to subtle to very subtle awareness, and further.

Diary after retreat, 2011

An observation: now that I've been involved in Mahamudra and

> Dzogchen insight practices, I'm reminded of the words of Lama Sopa that the immediate aim of Buddhist practice is: the perfect union of calm and insight, *shamatha vipashyana yuganaddha*. Alan invited us to do a little insight practice on mind, before diving into Settling the mind practice. When I do even a few minutes of insight practice before proceeding to Settling the mind shamatha, I experience better focus. Indeed, both Settling the mind and also breath practices get a boost, after I've first done some insight practice. Insight practice regarding what is mind, what is self ... feels like: familiarizing with impermanence, emptiness, insubstantiality. In Settling the mind shamatha observing, these aspects of impermanence and insubstantiality as to what arises and dissolves are so obvious. My experience is that this union of concentration and insight practice is working in a way of intensifying the effects of shamatha; of both practices ...

This connects with what I read in Daniel Brown's *Pointing out the Great Way*. He notes how he came to appreciate that the depth of realization that is possible during ordinary concentration and special insight meditation was strikingly enhanced by shifting to the very subtle or extraordinary level of mind. He explains, regarding concentration: in that sense, the issue became less about concentration on the object of meditation, and much more about the level of mind brought to the concentration.[24] As I understand his words, regarding concentration meditation: the process in shamatha gets enhanced when we are becoming able to practice, not so much from the coarse level of mind, but increasingly from subtle and very subtle levels. Indeed, *shamatha vipashyana yuganaddha* in various ways: consecutive, complementary, and in unity.

Expression, through the ages

Following are some additional text fragments by teachers in the Essential traditions through the ages, that have been clarifying for me, in seeing how mind practices – including Settling the mind – can be a foundation for, and part of the higher practices. They include a shift toward seeing that there is no substantial, inherent self that can serve as the point of observation during meditation. Moreover and more specifically: seeing that awareness-itself and ever-present awareness, instead of ordinary self-representation, get to serve as the vantage point for observing and knowing, during meditation. Mind and events (like thoughts, emotions, ideas) are nondual.

Tilopa, eleventh-century master, gives the following description, in the Mahamudra view: "Like the state of space, the mind transcends thought. Set it at ease in its own nature, without rejecting it or maintaining it. When the mind is without objective content, it is Mahamudra. Through familiarization with that, supreme enlightenment is achieved."[25] This is about mind transcending and including thought, mind and thought being nondual. As I learned, it is important to understand that phenomena will still arise and be present, and are needed for practicing in the Essential traditions. Yet, there is no grasping, there is a seeing through the ways the mind constructs reality.

Dakpo Tashi Namgyal (in the sixteenth century), speaking of the "abiding" and "moving" as the settled and dynamic aspects of the mind, notes that the meditator will "turn away from attachment and clinging to them as to separate phenomena. He will then cognize all the diverse appearances arising from the interaction of the senses and objects, including the mind's stable state and dynamic movements, as being the manifestation of the mind alone." [26] The mind alone ... abiding and moving being nondual, there is only mind, awareness, in this Mahamudra view.

Jamgön Kongtrul, nineteenth-century master practitioner in both Kagyu and Nyingma schools, and a pivotal figure and founder of the nonsectarian Rimé movement, states: "Whatever abiding or moving is perceived, it is unnecessary to fabricate anything ... Mahamudra, and common Dzogchen, whatever thoughts arise, without making anything out of them, you look nakedly right at them, and they become the path of liberation." In his commentary to

Kongtrul's text, Thrangu Rinpoche clarifies, as above, the nonduality of thought and mind: " ... if you have some recognition of your mind's nature, then, when any one of those thoughts arises, you will experience the mind's true nature in that thought, because the mind's nature is also the nature of that thought. It is the display of that mind. Therefore, when you realize that, whatever the initial content of that thought may be, it is liberated. It will not produce a second thought that's connected with it."[27]

In this brief exploration of wider contexts we have come a long way, starting from Settling the mind concentration practice on a coarse level of awareness. Then, a path was sketched by way of insight practices, to Essential realization of thought and mind as nondual, and "breaking through to pristine awareness." Then: all dust gone ...

Shifting to the blood and bones of practice again: here are some questions on my side. Connecting with Chapter 6, and my musings about the students of Lerab Lingpa and Düdjom Lingpa, I wonder: in what differential ways may the practices of shamatha, vipashyana, Mahamudra and Dzogchen have been presented and instructed, to many thousands of eager practitioners through the centuries? And in what ways might they most fruitfully be received, in the present time?

Ways of practicing, teaching and learning

The didactic ways in which teaching and meditation instruction are best given and best received will certainly vary, in certain time periods, in certain cultures. We can imagine that they will not be the same in, say, nineteenth-century nomadic Tibetan society, and twenty-first-century, often urbanized, so-called Western society. Imagine a nineteenth-century male novice sitting in the meditation hall of a large monastery in the Tibetan Himalayan hill slopes, together with hundreds of his co-students, living in a quiet pace. And imagine a Western student (maybe yourself) sitting in cozy but crowded urban housing, transformed into a meditation place; for one hour of practice, sandwiched between other activities, rush-hour traffic sounds in the background.

Daniel Brown, from his experiences as a Mahamudra teacher, names three areas where, in his view, Buddhist meditation teachings in the traditional forms they have been taught mostly in Tibet seem not so well matched to Western students. Connecting with some aspects that he names:

The first regards preparatory practices: in traditional meditation guidance students are taught a set of preparatory practices, designed to make the body and mind fit for meditation. They should make progress with meditation more likely (practices include the contemplations on the Four Thoughts that Turn the Mind, and the Four True Realities for the Spiritually Ennobled, Chapter 4.1). These preparatory practices are said to be not very popular with many Westerners as the form of presentation often is foreign to us. Quite a few Western practitioners are more likely to start immediately with formal meditation. According to Brown, these practitioners may recreate many of the bad habits of everyday living on the meditation pillow, with the result that they don't progress. Preparatory practices in some form are quite valuable.

The second regards methods of teaching. Looking at approaches that have been most effective in the West in bringing about real inner change: relationship-based methods of teaching seem best matched to Western culture. They seem to bring in better results than large group approaches with little direct contact of student with teacher. In most forms of psychotherapy the therapeutic relationship is crucial. Although psychotherapy alone does not accomplish the broad range of skills cultivated by the preparatory practices of the great contemplative traditions, Brown suggests that Western psychotherapy may be a culturally congruent way to accomplish at least some of the same objectives as the preparatory practices.

The third regards didactic approaches in meditation instruction: here Brown describes how within psychotherapy we have seen a cognitive-behavioral revolution over the past three decades. In these approaches, there is a targeting of specific behaviors, with detailed step-by-step instructions for change, and with clear descriptions of what can be seen as signs of progress. Present-day Westerners have learned to think in terms of step by step manuals. Brown, for this reason, believes that the Indo-Tibetan Buddhist gradual meditation stages approach to teaching meditation is well matched to Western thinking.[28]

Lama Palden, Western Vajrayana teacher and founder of Sukhasiddhi Foundation, notes that Westerners want to understand how other people did it, they like to learn through hearing what the process of others has been. How did you deal with that in your body, in your mind? This may include slowing things down, and exchanging about one's experiences. Easterners – again, generalizing – don't seem to talk much about this process aspect. [29]

So, some important aspects named here are the step-by-step learning, the personal relationship with the teacher, and sharing experiences. For me, step-by-step training, including sharing about the process, has worked as a powerful didactic tool, in various settings. I have enjoyed the step-by-step in the sense of the practices within the shamatha range we did with Alan Wallace: attention practices that have a sequential aspect, with body, mind, and awareness. Also the nine stages of the Elephant Path know their step-by-step instructions, pitfalls, and remedies. The Four Qualities of the Heart instructions, you may say, included an aspect of step-by-step in extending the circle of persons and phenomena to which we "sent" our deepening loving kindness. The shamatha teachings by Alan Wallace that I experienced in the Shamatha Project have been within – sometimes large – group setting, with sharing experiences. Next to that we had our fifteen minutes personal interview each week. Certainly, in Settling the mind practice, participants may be at varying stages as to their ability to experience the "space of mind," so group-teaching addresses various state-levels. Still, when we all practice shamatha, the range of possibilities in what practitioners experience is somewhat marked.

Step-by-step has some different feel for me in the Mahamudra and Dzogchen context. Here, in my experience, the range of attainment of the participants in a group, with shamatha, vipashyana and Essence practice, may be even much broader. Participants with varying meditation experience may do their practice from a different vantage point: more in the coarse or subtle or very subtle ranges. Here, general "one level" instructions can be confusing. I've enjoyed the privilege of attending week-long retreats in the Essential traditions both with large groups, and within small group. In the small group setting we received precise step-by-step teachings with detailed "pointing out" instructions, tailored for every specific person. The "pointing out" style refers to the original relational personalized style of teaching meditation, with rich details in descriptions.

These aspects, I suppose, may play an even more crucial role in year-long retreats. Practitioners in a traditional "three-year retreat," that includes Mahamudra or Dzogchen, are said to follow a sequential step-by-step approach, including these Essential practices in stages. While in a way they may be "progressing" as a group, the intensity of the personal process can be very challenging for Westerners as the path directly confronts and transforms habitual patterns and self perspective. Certainly here as well, the direct personal relationship with a teacher, with pointing out guidance, will be crucial.

Various styles, various settings, various aims and aspirations, various attainments, various realizations – all with their specific merits. With these remarks, while most of the previous chapters have been about the Shamatha Project retreats in 2007, I have come to more recent times. This is reflected in the diaries, that have been including musings about retreats in the sense of shamatha and Mahamudra-Dzogchen that I have participated in after 2007. On-going explorations.

And now: let's continue exploring shamatha while shifting the attention to this special and intriguing aspect in Settling the mind in its natural state meditation: insight into our psyche, its contents and its dynamics.

Chapter 8

Contemplative and psychological views

The reason we have this particular flat, human world – and the reason that we hang onto it so tenaciously and deny the existence of other possibilities – is because of what lies underneath it, above it, and outside of it. (Ray, 8.1)

Who am I, who are we, and: what is mind? What is referred to as its natural state? What you are reading is nourished by this quest with a mix of curiosity and yearning for understanding and freedom.

In earlier chapters I have described how the shamatha practices, and especially Settling the mind in its natural state, give us an opportunity to experience and observe, in a relatively safe laboratory situation, a wide range of ways and styles of dealing, coping, being with reality – within ourselves. In a sense, one would expect that everybody is always gravitating toward what is natural; that everybody, in some way is (on the way to) settling their mind "in the natural state." Basically this is the case, I dare to state. But how to find, reach, "live," "be" this natural state? In our zest for getting there we're up for quite some detours and turbulences.

The descriptions up till now have mainly been in terms of Buddhist meditation approaches, with a science view in the Shamatha Project, and personal experience from the inside – both from Tibetan Buddhist meditators and from me (Chapters 5 and 6). While Chapter 7 has addressed the larger Tibetan Buddhist practice picture, the coming chapters will explore further fields of interest connecting with shamatha, and meditation and science. Briefly, Chapter 8 connects with existential psychology, Chapter 9 with psychological turbulence, and Chapter 10 with science. These explorations relate to both Buddhist views and views from Western approaches, and connect with diary fragments. So, in the present chapter I include some psychological and existential lenses. I will look at ways of being with reality in terms of some views in Buddhist psychology, and of "Western" psychological approaches. These styles of viewing, and coping with the world, are connected with experiences as described in my diaries, and with descriptions by Tibetan meditators as presented in Chapters 5 and 6.

8.1 Various ways of being in the world

A few words about namings

Sometimes reference is made to Buddhist-contemplative and Western-psychological approaches. I cannot feel happy with this global addressing in generalizing divisions, there are some "apples and pears" in it. Also Buddhist-psychology and Western contemplation do exist: the combinations may be changed. Additionally, we know the term "Western Buddhism." Buddhism is connected with Dharma, a broad notion, that has been translated as standing for The Universal Law of Nature (the natural order of things), or Truth. Buddhism, in the Dharma sense, can be viewed as a science of mind, a religion-philosophy, a psychology, a blueprint for a civilization and a path of personal salvation. And Western psychology ... yes, is Western psychology, described as the scientific study of human (and other animal) mental functioning and behavior. Generally, I will use the labels Buddhist and Western-psychological, as these are the aspects that this writing focuses on. These considerations are included in what follows.

A Buddhist view: The Wheel of Life

Düdjom Lingpa refers to the wide range in human ways of dealing with meditative mind conditions, with: "The emergence, one after another, of all kinds of thoughts stemming from the mental afflictions of the five poisons, so that you must pursue them, as painful as that may be" *(13)*. This compulsiveness, during meditation practice, may also stand for a habitual approach. For these five, the often named three poisons: attachment, aversion, and ignorance, are complemented with pride and jealousy, two somewhat more complex states.

The Wheel of Life is an important image in the Buddhist world, and representations are found in most Tibetan Buddhist monasteries. Part of the imagery are Six Realms of Existence: hell world, world of hungry ghosts, world of animals, jealous gods, gods, and humans. I'll clarify what these stand for. For an example of the imagery, see Figure 2. While this figure shows additional meaningful imagery, in this context we'll just address the Six Realms.

Taken in a psychological way, it may be said that humans are familiar with all the five poisons, and all Six Realms may be recognized in the human psyche. Mark Epstein, psychiatrist and writer of many books on Buddhism and psychotherapy, gives an interesting description to the realms. According to him, this Buddhist picture of the Wheel of Life seems a particularly useful starting place in comparing Buddhist and Western notions of suffering and psychological health. What I like about the Six Realms imagery is that all realms show an inset of the *bodhisattva of compassion*, in a directly relevant form. This *bodhisattva* teaches us how to correct the misperceptions in each realm. The view is that, inherent in this image, the causes of suffering are also the means of release: it is our perspective that makes a realm into either a source of suffering, or a vehicle for liberation.[1] There is a way out.

The Six Realms can be seen as projections of our own mind as a result of our actions and intentions. They represent particular ways in which humans try to be real and solid, to exist. For some of us, a favorite realm may show in a striking character trait, for others all realms may present themselves in more fluid shifting states. I believe that, in principle, "nothing human is alien to any one of us," certainly that goes for me; as the diary descriptions show.

Connecting with texts and diaries

In Chapter 6 I gave some examples of parallels that struck me of the Tibetan descriptions of experiences a meditator may come across with my own. I can't resist the felt invitation to look at these six mental states as described in the age-old Wheel of Life. Can the experiences of a twenty-first-century Western woman meditator

Figure 2: The Six Realms in the Wheel of Life

connect with a Buddhist typology? Can the experiences of a nineteenth-century Tibetan meditator join with a Western typology? With a direct question: do I recognize my diary descriptions in these states? For instance, did I visit hell?

After Figure 2 a description is presented of the way these Six Realms can be experienced. Then you find some examples for, shorthand, the Buddhist phenomenological descriptions in the texts by Lerab Lingpa and Düdjom Lingpa. Both teachers will have been brought up with the Six Realms as vital perspectives. These descriptions are followed by my (Western inspired) diary fragments.

1. Hell

Hell imagery, the lower-middle realm in the image, shows a place of tremendous suffering and torment, with fires, and images of beings crying in distress. There is intense inescapable searing cold and claustrophobic heat. This is the world of utter aggression, terror, fear.

In the texts by Lerab Lingpa and Düdjom Lingpa, a hellish reference by Düdjom Lingpa can be found in a description by his students of "The feeling that gods or demons are actually carrying away your head, limbs, and vital organs, leaving behind only a vapor trail" *(27)*.

Regarding the diaries: there have not been any really hellish experiences on my path. Some experiences have been more pleasant than others. Although hell has relatively spared me, for the sake of an example: diaries show in (15) a micro-hellish sequence (a rather benign variation). This shows in the 1. desire for the other person's chocolate bars, the 2. "burning" envy toward the person who received them: I'll pick it away, don't want her to have it, don't like her; then the 3. inner response: oh, I don't like feeling this burning envy, 4. but it's there, all inclusive, acceptance ... and 5. the witness: more acceptance, soothing. Diary descriptions also contain possibly hellish references to temperature, like (10) hot flushes, and (33): cold, feeling broken, like experiencing "cold turkey," as I imagine that to be.

The Buddhist approach stands for non-denial, as is also clear in the *bodhisattva*, who carries a mirror, showing that this suffering can only be responded to by seeing oneself with the unwanted emotions in the mirror. The need to be free of destructive emotions is recognized, while at the same time it is acknowledged that such freedom can come only by way of non-judgmental awareness of just those emotions of which we want to be free. One of the contributions of the Buddhist approach is its ability to teach effective methods of relating to one's own destructive emotions.

2. *Hungry ghost*

The lower-left image shows creatures with shriveled limbs, gross big bellies, and long thin necks. This is the realm of intense, insatiable hunger, thirst, neediness; the world of desire combined with rage, and poverty mentality. Next to the neediness, the painful thing is that, with their thin necks, these beings cannot take in the food that may be found, because attempts to ingest cause even more pain. The *bodhisattva* of compassion often carries a bowl with objects for spiritual nourishment (here we are talking about a different dimension than envy toward the one who has the chocolate). A Western counterpart may be found in the psychodynamic notion of "oral craving," a sense of very intense obsessional neediness.

Epstein makes the interesting association here of this thirst, and neediness, with the phenomenon of low self esteem and feeling of unworthiness in many Westerners, including Western Buddhist students. This neediness has proven very difficult to understand for Eastern teachers, who are not familiar with this phenomenon.[2] Related feelings may show as a sense of insecurity, desperate longing for "outside" confirmation and support from the teacher. Various kinds of so-called transference phenomena (standing for a displacement of wishes and feelings related to persons from the past, now toward the meditation teacher) may be relatively benign-in-passing or quite pervasive. They play some part in any retreat, our expedition not excluded.

In the texts I find Düdjom Lingpa mentioning, in *(8)*: Compulsive hope in medical treatment, divinations and astrology; yes, it has to come from outside.

As to the diaries, I can give a – benign, again – example: the weekly interview with Alan Wallace is about insights, exuberance, thrill. I mention "processes of anxiety, feeling of loss of control," there is a sense of insecurity, and a wondering: "wishing for his approval?" – thus I mention (26).

3. *Animal*

Other than beings in the lower realms of hell and hungry ghosts, animals, in the lower-right image, are part of our everyday experience. We can see, hear, touch them and interact with them. Animals undergo deep suffering, eating and being eaten, hunting and being hunted by each other and by humans who, in their self-centeredness, add a lot to animals' suffering. Animals have a relative fixity in habitual patterns, dictated by the limitations of their nervous systems and physical bodies; and animal psychology cannot relate to ambiguity, newness and uncertainty.[3] The animal realm also regards drives, instincts, and sexuality. Sexuality by itself is not the problem, but while there are, on the one hand, inhibitions around freedom and happiness in sexuality, on the other hand there is indulgence, by an attempt to get lasting pleasure from what is unavoidably transient. The *bodhisattva* of compassion here is holding a book, pointing to the capacity for thought, speech and reflection that our animal natures lack.

For the texts: in a sensitive way, Düdjom Lingpa refers to "Grief and disorientation when you wake up, like a camel that has lost her beloved calf" *(11)*.

Regarding some animal experiences in the human realm, I find in the diaries: (34), a reference to a memory of being on the beach with a lover, exuberantly making love, excitement, indulgence. I describe some energetic strain, letting go of the grasping, still the fantasy arises, is, dissolves, things are stirring. In the middle of experience not much reflection is found.

4. *Jealous god*

The jealous gods (also called half-gods, upper-right in the imagery) are separated from the god realm by a wishing-tree full of fruits; they are quarrelling among themselves, and always jealous of the gods who have more, or better, and are superior. This is the realm of ambition, jealousy, and paranoia. Reginald Ray, Buddhist scholar and Vajra Master, describes paranoia as a kind of radar system, the most efficient radar that the ego could have. While this radar picks up all sorts of faint and small objects and impressions, suspecting each one of them, every experience in life is felt as something threatening. The *bodhisattva* of compassion on the image holds the flaming sword of discriminating awareness. The presence of this sword underscores the point that, by itself, the aggressive nature of the ego-personality is not seen as the problem: this energy is in fact valued and needed on the spiritual path. The problem is in the objects of aggression: the *bodhisattva* is urging the jealous gods "to redirect their aggression, destroying and assimilating the unawareness that keeps them estranged from themselves."[4]

In the texts we find Düdjom Lingpa mentioning "Unbearable anger due to the paranoid thoughts that everyone is gossiping about you and disparaging you" *(18)*.

The diaries give some examples of this checking radar: I'm so aware of these micro operations of habitually locating myself in checking coordinates about where, what, who I am. I wonder about an underlying: anxiety, so habitual, so craft-fully covered, in (20).

5. *God*

The god realm, upper-middle, refers to experiences of pride, delusion and addiction to on-going pleasure. The *bodhisattva* of compassion appears here holding a lute, signifying musical enjoyments, but, as it is said, also alerting the ones in this realm to the

Buddha's teachings. This realm, still part of the on-going cycling conditioning of samsara, has the problem that the state of mind is one of intoxication and pride. This gives the inhabitants the handicap that precisely because of their power and longevity they are not able to hear the Dharma. They can't hear the teachings on *duhkha,* suffering, reactivity, and the ending thereof, as explained in the Four True Realities for the Spiritually Ennobled (see Chapter 1.6.). This can be the realm of, what are named "peak experiences," said to refer to certain ecstatic mental states that are connected with a sense of harmony, euphoria, and interconnectedness. They are in this realm, if grasped at. These experiences can happen both during meditation and outside of it. The god realm may also relate to "being in flow," like in performing sports, or painting, and also in meditation. Being in flow has been described as a state in which a person is fully immersed in an activity, with a feeling of energy, focus, and full involvement, and with the experience of the activity as intrinsically rewarding. [5] The problem, again, comes in when the person is grasping, and gets attached and addicted to these temporary states. Regarding meditation: these are powerful states, cultivated in Buddhist practices but also warned against in the teachings, as they may induce an inner complacency, and take away the aspiration to continue on the path. This regards the *jhanas*, the deeper concentrative absorption states. Alan Wallace refers to this kind of experience also in relation to the attainment of shamatha; and, nearer to experience, to reaching the fourth of the nine stages, when continuity in attention is reached. Some glimpse of a possible complacency I recognize: I described this in diary (16), "so happy I got here!"

As to the texts: Düdjom Lingpa mentions that stillness is pleasurable, but movement is painful *(3).* This may of course be a momentary description. The problem comes in when attachment to stillness brings in fixation, trance, and a halting of the process.

From the diaries comes: a feeling of expanding into a larger space, with a flash of grandiosity, and the sense that then the whole universe is "me," with a seductive power in it (47).

6. *Human*

"The preciousness of human life!" As was mentioned before, the human birth is considered the most auspicious one for spiritual development. In this realm, upper-left in the imagery, enough pain is present to motivate one to practice on the path. However, the pain is not of the degree that one is completely preoccupied with suffering and paralyzed by it. The psychology of the human realm is about desire and discriminating passion. Hope and fear play an important part and take us outside of the moment, to future and past; urging for constant discursive thinking and strategizing. The *bodhisattva* of compassion appears in the form of the historical Buddha Shakyamuni, with the staff of an ascetic and the alms bowl, committed to the basic Buddhist quest for seeing the truth of who we are, without distortions. So, in principle we have the capacity of seeing and knowing, based in this fundamental form of reflexivity as an implicit aspect of all cognition, addressed in Chapter 7.3. Yet, hope and fear and underlying self-attachment draw us easily back into more "animal realm"-like irritability and reactivity aspects.

Interestingly, within this realm, of the huge range of possible states and ways of being, by far most of us only get to know a tiny fraction during our lifetime. Reason is that we keep ourselves locked up in what has been conditioned into a feeling of normality. This is a normality that is blind to its own narrow range, and to its own distortions of reality, a quasi normality for humanity that cannot jump over its own limitations.

The human realm is thoroughly interrelated with all other realms. I feel struck by Ray's words, stating that: " ... there are experiences of the other realms within the human one, windows to the other realms, and even glimpses of spheres that are entirely outside of samsara. Our subliminal awareness of these other possibilities of existence undergirds our human world, contributes to it, and helps make it what it is. In other words, our flat world cannot be explained on its own terms." And he continues: "The reason we have this particular flat, human world – and the reason that we hang onto it so tenaciously and deny the existence of other possibilities – is because of what lies underneath it, above it, and outside of it. To put it in a nutshell, maintenance of our human realm depends on a large-scale denial of reality as a whole."[6]

As part of this denial of reality we see that in the human realm the tendency to hide from ourselves is most clear. Why? What makes us so afraid of looking at ourselves, or being really "seen" by others? Why are we seemingly defending so fiercely, not wanting to be "discovered" ... One would surmise that in this case the thing that is feared must be really terrible. What makes us feel so vulnerable as we approach our deeply personal and private feelings, with this question of "Who am I?" Taken from the picture that arises from diaries, from texts by Lerab Lingpa and Düdjom Lingpa: it is clear that anxiety, fear, terror here are forefront. There is the fear at discovering our own insubstantiality, embedded in conceptualizations that this discovery means death – which is true, in some sense. It is true, from the perspective of ego-personality. The good news is that Dharma practice enables us to genuine liberation by truth, rather than permanent isolation in hopes and fears.

As to the texts, indeed, Düdjom Lingpa refers to our human condition as characterized with "all kind of hopes and fears" *(22)*.

In the diaries, it is mentioned that then, after the agitations, defenses for survival, there is a shift to feeling real powerlessness, just feeling sad, and then a sense of great relief, joy, sense of freedom, "just this" (30). At that moment: truly human, at last ...

Overall, this manner of looking at the Wheel of Life presents various ways in which beings, among them we humans, are engaged in maintaining themselves. Also, it clarifies the repetitive, self-defeating quality of each of the psychologies involved. No lasting peace to be found here.

Compassion for ourselves doing the practices – guided meditation

Before continuing this exploration into a Western view on psychological states, let's first do a meditation practice that connects with this precious human condition. The human condition refers to the only realm from which lasting escape from suffering is possible. In this human condition we may still have such a hard time seeing the reality of who we are; including during meditation. The following meditation guidance was presented at a moment that many of us at the starting phase of a group retreat experienced quite some doubts and struggles in doing the shamatha practices. Let us invite all *bodhisattva*s of compassion and know them to be outside *and* inside of us. This was a guidance that I was allowed to present to a small group on a shamatha retreat in Phuket, Thailand, home base for Alan Wallace's retreats since around 2010.[7]

> Let's settle the body in its natural state, imbued with the qualities of relaxation, stillness and vigilance. Find your posture of comfort and ease and, at the same time, wakefulness. Be aware of how you sit or how you lie down supine, where the body touches the ground, the chair, the cushion. Sit-bones, backside of the body, spine ... Perform a brief body scan while relaxing and releasing any tension, let it melt ... Take delight in a few deep breaths, deep out-breaths especially, with on the out-breath letting dissolve any contractions, subtle little knots that still may be there. And let the breath find its own rhythm, and allow awareness to suffuse the whole body, while you are aware of the tactile sensations of the breath moving throughout the body. And now, while keeping a peripheral awareness of the breathing, the oscillations, shift the mind in its dynamic, creative, imaginative mode.
>
> Compassion: the heartfelt yearning that all, including yourself, be free from suffering and the causes of suffering. With that goes a wish to be of help. Combine this yearning with this

deep wish and longing: may I, like all other beings, be happy and know the causes of happiness. Now, let's zoom in to ourselves, doing the practices. There may be joy, and sometimes there may be struggles. We may express this yearning: may we be free of the struggles, the suffering going with the struggles that hinder the meditations. Some effort-stress may be all in the game, but there is extra suffering that we put to it, like with these ruminations. May I be free of these endless, repetitive, self-centered ruminations, hindering the practice. May I be free of the causes of these unwholesome ruminations in grasping and in anxiety, and may I be free of the judgments about this all, and about my coping, when it goes in unwholesome directions.

May I have compassion for myself, may I be of help to myself. May I be open to the knowingness that I'm worthy of this compassion, may I trust. And, as Alan explained this synergy of the Four Qualities of the Heart: when compassion has a hard time, when the stallion of compassion stumbles, there are these guardians, or remedies: the first remedy for compassion is Empathetic Joy. Next to the constriction felt-sense of the ruminations: may I take delight in doing the practices. May I congratulate myself for having made this commitment for being here, for dedicating these months to Dharma. The second remedy is Loving Kindness: may I be happy doing the practices, may I know and fully be aware of the gift, the wholesomeness, the enhanced sanity that these practices are bringing.

And now expand: may we all be happy and know the causes of happiness: you, me, all who are on a path of practice, on a path of sanity, a path of wisdom and truth; with compassion for ourselves, when struggle arises. And expand further to all who are worthy of compassion: as all are! May we all be happy, no one excluded, and know the sources of happiness – and may we all be free of suffering and know the causes of suffering.

Let's continue practicing in silence … And let's conclude this session.

I'm so aware, when writing this down, of the different context in which these instructions were given, as compared to the context while in the Shamatha Project. Tropical Thailand, on the one hand, and the icy radiant Rocky Mountain atmosphere, on the other! Still, with comparable meditation instructions and experiences (Experiences perspective). The present instructions bring me back to felt-sense memories of Thanyapura Mind Centre, this "Land of Abundance" center in Thailand where Alan Wallace offers eight-week shamatha retreats. Felt senses: view of palm trees, deep red sunsets that suddenly shift into darkness, massive undulating sound of cricket choirs. The gentle morning sun warmth on the skin at an early walk, with moist jungle smell … and then, at breakfast the feast of the first bite-taste of dripping golden ripened mango pieces (such a different context in terms of Cultures perspective). While no science measurements are made here, still our neurophysiology will show its dynamic correlates (Behaviors perspective). Same meditation, within such a different geopolitical situation, organizational structure, and ecological niche (Social Systems perspective).

A Western-psychological view: the Diagnostic Statistical Manual

After the Buddhist way of "mapping" human realms of existence in six possible states of being, what can be said about the ways that Western psychology and psychiatry view people in their ways of dealing with reality? Globally, here we are talking about the way that human persons and personalities present themselves to others and themselves. The term personality has generally been used, in Western sense, as referring to an individual's characteristic pattern of affect, emotion regulation, behavior, motivation, cognition about self, and interactions with others that are long-standing, generally present since adolescence. For the West, the term "personality" may carry certain qualifying adjectives, general qualifications like ambitious, or easy-going. Also the qualification has received psychiatric significance, like when talking about a passive or aggressive personality. As to psychiatry, various "personality disorders" are described in the Diagnostic Statistical Manual (DSM), a Western manual for psychiatric categorization. [8] This categorization is based in statistical analysis of symptom clusters as they showed up in field trials. Benefit of a categorical approach can be that it provides a common vocabulary for mental health professionals, as to presenting symptoms in a person they are working with. With certain symptoms, a person is said to have a certain disease. Another approach sees most human behaviors and symptoms as rather occurring on various continua. Starting from these continua is referred to as a dimensional approach to diagnosis. With a dimensional approach, the emphasis is more with a person that can have some measure of various ways of suffering, illness, and health.

Looking at diaries and texts again: certainly, quite a variety of anxiety expressions came by. In both diaries and texts, passing auditive hallucinations have been mentioned, like my hearing sounds that were interpreted as if the building would be besieged (31), and for Düdjom Lingpa's student the sense that there is some special meaning in every external sound *(16)*. Also, reference has been made to flashes of paranoia, like in (30), about the sense of things done purposefully against me, and *(7)*, an inexplicable sense of paranoia about meeting other people. Well, as temporary *nyam*, they will not have been "enough" for counting for DSM categorization; yet ... We'll be back at the theme of discernment and differentiation in this area in Chapter 9.

8.2 Musings: Buddhist and Western-psychological manifestations

The named Tibetan categorization, at the least, can be said to impress as quite imaginative and colorful. Regarding neurotic suffering – as one class of possible suffering, in a wide range – Epstein suggests the following East-West bridge. He notes that different schools and movements in Western psychology have done much to illuminate the Six Realms dynamics. Next to elaborating on more wholesome aspects in the mind, Freud and his psychodynamic followers insisted on exposing the animal nature of the passions, the hellish nature of aggressive and anxiety states, and the insatiable longing of what came to be called "oral craving." In most ways of picturing the Six Realms, these named aspects refer to the lower three realms. Later developments in psychotherapy shift the focus to the three upper realms. Ego psychology, behaviorism, and cognitive therapy cultivated the competitive and efficient ego-personality, as seen in the realm of the jealous gods. Humanistic psychotherapy emphasized the "peak experiences" that may go with the god realms. And, the psychology of narcissism can be seen as addressing specifically the questions of identity, of a more wholesome or unwholesome sense of self, so essential to the human realm. Each of these approaches in Western psychotherapy, here presented in a bit of a caricature, has had important contributions in the understanding of who, what and how we are. In the words of Epstein: "Each of these trends in psychotherapy was concerned with returning a missing piece of the human experience, restoring a bit of the neurotic mind from which we had become estranged." So, combining these views may invite us to be somewhat less estranged from our neurotic mind. Well, anyhow: still neurotic! Fundamental to the Buddhist notion of the Six Realms, neurosis, and worldview in general is the following: it's not only that we are estranged just from these aspects in our character. As the Buddhist teachings assert: we are estranged from our own Buddha-nature, from our enlightened minds.[9]

Here are some more musings going with the named Buddhist and Western typologies that reflect the fundamentals. While the examples of the Six Realms and DSM cannot really be compared, they connect with some interesting cultural and existential themes, I believe.

Regarding the Buddhist example: as mentioned in 8.1, each of The Six Realms may show in a person, as connected with striking character *traits* (more in the sense of relatively stable enduring traits), while also each of the Six Realms may be connected with more fluid shifting emotional *states*. Both possibilities are included. The Buddhist view, generally, emphasizes impermanence, and in that: fluid, shifting states. States will change. Traits, where repetitive states have become more enduring, can still change with the appropriate practices. Also, the Buddhist typology may be seen as more of a dimensional approach: more or less of this or that, in varying degrees that change overtime. Many of us know some hellish experience, as well as some "god-like" addiction to on-going pleasure.

Regarding the Western example: DSM provides a system for categorizing psychopathology, it addresses what are seen as deviant, abnormal, and maladaptive behavioral and emotional conditions; this is where the emphasis lies. While this is a different entrance than The Six Realms, and in spite of this "apples and pears" aspect, still, may I address a few tentative global observations. Terms like deviant and abnormal point to an imagined "normality" from which one may deviate. While this is all about humans, this sense of deviant or abnormal for many of us (who take themselves for normal) seems to support the view of placing this condition with "others," in that way reflecting and supporting the separate-self sense; and vice versa. There seems to be a suggestion of not only "other than normal," but also: other than we. In comparison to that, the Six Realms appear to give a more inclusive feel, relating to "all of us." Regarding fluidity: certain states and traits can be addressed with certain kinds of therapies, change is possible. While some states and traits are more dimensionally conceived, categorizing doesn't have a felt-sense of fluidity.

DSM is categorizing and labeling, in a Western-psychological context, in a way that many "idioms of distress" in other cultures are missed. And by the way, categorizing – other than what "diagnostic" generally includes, in some sense – doesn't give any indication about personal developmental history, neurobiological substrate, context, or optimal therapeutical intervention.

After addressing these examples of typologies, let's now look from a different perspective.

8.3 Defensive patterns in workings of the mind: avoiding what's here

Who am I, what is mind? What is the natural state? Let's start now with a Western contribution to seeing people in their diversities. In this, we look at what may be considered specific repeating states that are recognized in so-called "defensive styles," connecting with traits. People use these styles in dealing with the world including themselves, and they often have their favorite styles. After a presentation of a Western styles approach, an example of a Buddhist view regarding defensive style is given.

In this Western context we look specifically to seven levels of defensive functioning, that are seen to characterize a person in important ways. This is a very helpful and practical contribution to our understanding in the sense of development, and stages. What needs to be defended against? What is in need of being defended? We touched upon that question, in addressing the human realm of the six. The title of a recent book on defense mechanisms leaves no doubt about that: "Protecting the self." As it is said, the function of the defense mechanisms is twofold: to protect the individual from experiencing excessive anxiety, and to protect the integration of the self. In this Western view, these are crucial functions. Anxiety is seen as the driving force behind psychopathology. Anxiety, according to psychologist Leigh McCullough Vaillant, connects with inhibitory affects that signal conflict and lead to defensiveness. These affects are seen as part of our biological endowment. While the underlying mental mechanisms are biologically prepared, the ways they are behaviorally implemented are acquired. Defensive behavior patterns, to the extent that they have been learned, can be unlearned, says McCullough Vaillant.[10]

About the use of terminology

Terms like "personality," "ego," "self" and "separate-self sense" have very diverse connotations in various cultures and subcultures, so I like to bring some clarity about how I use them here.

The term *personality,* referring in Western context to an individual's characteristic pattern of affect, emotion regulation, behavior, motivation, cognition about self and interactions with others, has been used as a global way to describe and label a person's observable behavior. Next to that, it includes the person's subjectively reportable inner experience. I would like to note here the aspect that this description carries the sense of the person being identified with this designation and labeling of personality, independent of content. This is not mentioned in the description, as it is considered self-evident. Personality, also referred to as ego-personality or character, can be conceived as one's typical, habitual identification and conditioned pattern of adaptation.

Ego, in Western psychodynamic context, is seen as the aspect of the psyche that in its functioning controls movement, perception and contact with reality. Moreover, through the mechanisms of defense available to it, it controls the delay and variations in the way that drives are expressed. *Ego functions* are considered to include judgment and the capacity to form relationships. Also ego is seen to have a capacity to integrate diverse elements into an overall unity. In this Western view, effective ego-defenses are "working" when they can abolish anxiety and depression. On the other hand, when we abandon a defense or when a defense "melts," this increases our conscious anxiety or depression, and may lead to reactivity and aggression.

Self in Western psychological sense globally refers to a person in the light of the person's own individuality: it is the individual person from her or his own subjective perspective. Self, experientially: that's me. For Western psychology, there is much emphasis on individuality, autonomy, independence, and the self experience as separate. This is often translated into the felt sense of self as being a separate entity. Abraham Maslow, known as a leader in humanistic psychology, proposed a hierarchical organization of needs present in everyone. First the more primal needs, such as hunger and thirst, have to be satisfied. Then more advanced psychological needs, like affection and self-esteem, become the primary motivators. Self-actualization is often quoted as being the highest need that Maslow proposed. However, later in his life in publications starting from 1969, Maslow emphatically added the higher need of self-transcendence that includes seeking to further a cause beyond the self, like service to others.[11]

Looking from a wisdom traditions perspective, Western psychology is dealing with an objectified, reified variety of self experience, taking that for real. By the way, for Buddhism, as we saw, we can certainly speak of self experience, and self sense; however, with personal research, an inherent independent separate-self cannot be found (Chapter 2.6). It is in that sense, that the expressions "non-self" and "emptiness of self" are used, referring to this self as non-separate, functioning in a matrix of interdependence.

In the terminology that I use, we do need ego in its functioning, for our functioning. For walking, for seeing, for thinking, for being in time, for writing this sentence, for reading it and turning the page ... ego is involved in all of our functioning; in this view: ego functioning is a bare necessity, let's be grateful for that. Next to ego in this more fluid sense of "functioning," I use the term personality in the sense of the person identified with, and attached to her or his habitual tendencies and patterns. This identifying attachment limits us in our potential. Some identification may be part of our development; and in the process of transcending and including, it's good to realize that this, too, belongs to the inclusiveness of who we are – and in that sense, deserves our love. This personality, or ego-personality aspect in some spiritual circles is referred to as "ego" – which easily leads to confusion of tongues.

When I use the term *self,* this is mostly in the general way of the person from her own perspective, and the context will make clear if the sense of self has a more or less attached way.

Next to this, I use the expressions *separate-self* and *separate-self sense.* While ego-personality and separate-self refer to a *concept*, separate-self sense refers to the felt sense of the self, taken as separate – the felt sense of the attached ego-personality, which many of us will know, and be aware of, more or less consciously.

A Western-psychological view: seven levels of defensive functioning

"Defense mechanisms," in the Western view, can be seen as mental maneuvers in which we all engage to maintain our psychological equilibrium and protect our sense of self. They are the unconscious mental processes that ego uses to resolve conflicts among four important aspects in inner life: drive, (instinct, need, wish), reality, important persons, and conscience. Defense mechanisms are distinguished from coping mechanisms in that coping involves a conscious, purposeful effort, with the intent of managing or solving a problem situation, whereas defense mechanisms occur without conscious intentionality. Defenses function to change an internal psychological state, but may have no effect on external reality and thus may result in reality distortion. So, what seems to be "protecting" and balancing, also supports distorted perception. In that way, it supports ignorance and suffering.

Defensive patterns have been categorized in various ways. Here I'll present the "defensive functioning scale" of the Diagnostic and Statistical Manual of Mental Disorders, fourth edition, DSM-IV-TR with text revisions. I'll mention 31 forms of defense, allocated to seven levels that run from high to low adaptation. [12]

Interestingly, it turns out not to be that difficult to recognize all levels, from mature to very immature, primitive narcissistic and psychotic, in the experiences that I describe in the diaries ... *nyam* of all kinds, that can be viewed in a dimensional way. Certainly as interesting: the phenomenological descriptions by Lerab Lingpa and Düdjom Lingpa can very well connect with these conventional

Western psychological notions as well. So, this is the other way round: as it was possible to recognize Western-framed experiences within the Tibetan Buddhist Six Realms (8.1), also the Tibetan descriptions can ally with present time Western defensive styles.

Following are the level-descriptions from DSM-IV-TR, accompanied by illustrative examples from diary fragments regarding Settling the mind in its natural state, and by fragments from the texts by Lerab Lingpa and Düdjom Lingpa.

1. High adaptive level

Functioning on this level stands for optimal adaptation in the handling of stressors. Some examples of this way of functioning are: anticipation, using humor, self-assertion, and self-observation. Isn't that interesting, to call these wonderful phenomena "defenses"! Defending against what, and what needs to be defended?

Regarding the diaries: certainly, this "self-observation" is nourished by, and included in Settling the mind in its natural state practice. Incidentally: the Settling practice covers a far larger notion of self than is addressed here! For an example in the diary: I observe myself feeling really powerless and just sad, and then there is a sense of great relief, joy, sense of freedom. "Just this" (30).

When I read the description of high adaptive level, and the texts, then I get the feeling that the practice of Settling the mind, as known in Buddhism, may be seen as a beautiful universal tool for cultivating high adaptation! Self-observation: a defense? Or: reflexivity following the thread of truth? What touches me as in the same atmosphere of high adaptation, is Lerab Lingpa's way of trusting to you, the meditator, to "act as your own mentor;" and Düdjom Lingpa's mentioning of the highly adaptive realization, respectively motivation, of "disillusionment with samsara, and your heartfelt compassion for sentient beings" *(25).*

2. Mental inhibition or compromise formation level

Here defensive functioning keeps potentially threatening thoughts, memories, feelings, wishes, or fears out of awareness. Some examples are intellectualization and isolation of affect; for instance by telling about an emotional experience in an intellectual insensitive way.

In the diaries I recognize these maneuvers in the tendency, certainly in the beginning, of connecting experiences to familiar phenomena "at home," comparing, translating, conceptualizing and intellectualizing. Some associations that in the early weeks came up with compulsive force have appeared under "next." One example is my intellectualizing about EEG patterns, beta, alpha, theta and delta rhythm, how about gamma (25), in the context of a dazzling week, including "processes of anxiety, feeling of loss of control" (26). Intellectual activity can be fascinating and exciting; however, intellectualizing may foreclose other more emotional or existential material. Maneuvers on this level are often much accepted as normal, in our culture.

As for the texts an association can be made to Lerab Lingpa's first sentence: "Simply hearing your spiritual mentor's practical instructions and knowing how to explain them to others does not liberate your own mind-stream, so you must meditate." Meaning: you can't hide behind the intellectual stuff, you have to live it, "warts and all," including all unexpected emotions that may come up!

3. *Minor image-distorting level*

This is about distortions in the image of self, body, or others that may be employed to regulate one's self-esteem. Examples include devaluation or idealization of others and oneself. Idealization of oneself might show in boasting about successes, while maybe unconsciously feeling insecure.

As for the diaries: I find a touch of "boasting," with the risk of complacency, in more stability in not losing the thread of continuous mindfulness. So happy that "I got there!" (16). Complacency ... mmm, in my memory there also was a sense of great relief and spaciousness, however relative. The problem lies in the attachment.

This resonates with Lerab Lingpa's admonishment in his text: "It does only harm ... to claim, 'I saw a deity. I saw a ghost. I know this. I've realized that' and so on." This defensive boasting may temporarily seem to help for an uncertain meditator, but the effect is fixation and less openness in on-going development and unfolding.

Reading these examples now and writing them down: I like to add that, of course, with all these examples, there may be a touch of something – some defense, some "wart" of attachment may play a role; while at the same time, a feeling can be quite authentic, and mixed with joy.

4. Disavowal level

At this level unpleasant or unacceptable stressors, impulses, ideas, or feelings are kept out of awareness with or without misattributing these to external causes. Examples are denial and projection. For instance, like someone who does not recognize her own jealousy but ascribes this, projects this to another person: it's not me but you who is jealous! Projection often goes together with a feeling of aversion, as that which is projected is felt as unacceptable.

For the diaries: compared to the first three levels, here at level four I have to look for examples where denial and distortion are getting more robust, and more hidden in unconscious recesses. I recognize a relatively benign example, about these delicious chocolate bars, again ... desire, inner taste. With a flash of denial, "don't want to feel this longing," followed by the envy and the tendency to "disavow" this unaccepted feeling (15).

In Düdjom Lingpa's text the following is mentioned: "Fear and terror about weapons and even your own friends, because your mind is filled with a constant stream of anxieties" *(21)*. In the examples at the High adaptive and Mental inhibition levels, a faster recovery was possible, there was more fluidity in the maneuver. Here's a variation where it may take some more time: there's anxiety, and probably anger, leading to projection of inner threats to outside, to others. Noteworthy is that here it becomes less easy to discern these patterns in oneself, in myself. This is so because with more coarseness, self-observation and self-reflection will decrease (so, in principle, when in the coming sections I say: I haven't found this or that maneuver, I may not have been able to "see" it, let alone to describe it).

5. Major image-distorting level

At this level, gross distortion or misattribution of the image of self or others is in place. For instance, there is splitting in the self-image or image of others, a splitting in good versus bad extremes. This may be in the form of putting the "bad" rage away by projection to another person: "it's his anger, not mine." With so-called projective identification, there is an active attempt to keep some control about what's projected. This may feel for the other person as being sucked in the dynamic.

Diaries: this gets even more serious. An example of subtle splitting I give in the distortion, devaluation in self-image: ridiculing, accusing turns toward myself: not able to surrender, always doing things wrong, in (30). Here the "always" is a warning for generalization, which often is part of splitting. Note that in a feeling of utter "badness" there can be a flavor of glorious grandiosity. Nobody as bad as me! Which in a paradoxical way, in this kind of situation, may feel "good."

The texts offer: "Unbearable anger due to the paranoid thoughts that everyone is gossiping about you and disparaging you," as Düdjom Lingpa mentions *(18)*. Here no quick recovery can be expected, because here anxiety has found its covering counterpart in anger, and is rigidified in a paranoid identity. Crucial is that in this case there's not a mentioning of "as if" aspect, like there may be in somewhat more mature levels. The person is really identified with the perception. In "everyone else" you find the generalization again. There is projection of devaluation to everyone else doing this about you.

6. Action level

This way of defensive functioning deals with stressors, internal or external, by action or withdrawal. Examples can be found in "acting out" and passive aggression.

In the diaries, in example (32), I name the tendency to underreacting (inner reactivity of repression, suppression), that may have the withdrawal aspect, while the overreacting involves the tendency to act out, to do things in an impulsive way. Both may lead to some very temporary seeming release.

In the texts I find: "Negative reactions when you hear and see others joking around and laughing, thinking that they are making fun of you, and retaliating verbally" *(19)*. Here the retaliating verbally stands for the acting out, impulsively, without intermediate feeling and thinking, that might modulate the tendency to react in this way.

7. Defensive dysregulation level

At this level, defensive regulation fails to contain the person's reaction to stressors, which leads to a pronounced break with objective reality. This may show as a psychotic distortion, a delusion or delusional projection, disconnected to reality. In common language usually referred to as: being mad, insane, or crazy.

A diary example presents me hearing a horn blowing, like: up to the fight. I go outside, hear nothing and suppose that it may just be some uneven noise in the heating system. Here we can speak of a brief hallucination, a moment of psychotic distortion. The moment I go outside I'm exploring, and then I can correct my brief mistaken perception (31).

In the texts there is a description of *(27)* "The feeling that gods or demons are actually carrying away your head, limbs, and vital organs, leaving behind only a vapor trail ..." The "actually" may be stressed here: it's for real, for the person. Others may see it as impossible, a psychotic distortion.

Diary after retreat, 2009

> Connecting all these has been quite a job, and in a way it is confronting. It is all descriptive, and dimensional (and not categorical): somewhat more, somewhat less of this or that. I have experienced the composing of this little matrix a humbling, and at the same time liberating endeavor. A real equanimity practice with seeing the fluidity of all these qualities and states, and not being identified with or attached to any of them. With the beauty of the Settling the mind self-observation, and the self-reflection, that enable me to look in this way.

We saw that in the mentioning of Maslow's needs, the highest level of self-transcendence is often not included. Probably even less we could expect any mention of "higher," more and more mature levels and styles of defense and coping, in a transcendent sense. Indeed, they are "not needed" when the scientific approach does not include higher dimensions of consciousness. Still, they have been described in the transpersonal sense, characteristic of, especially, the subtle and very subtle realms. Ken Wilber and Roger Walsh, among others, are describing these in detail. May we name these "defenses": Level zero? One example is "psychic inflation," which happens when universal-transpersonal energies and insights are exclusively applied to the individual self, with unbalancing effects. A second example is the trap of "spiritual bypass": when a person uses spirituality to deny or avoid psychological issues. For instance, a meditator who is uncomfortable with intimacy might rationalize her avoiding of relationships as "wishing to live a solitary spiritual life." Other examples include "metadefenses," expressions of "defending" against knowing and growing to higher states and stages, as these may represent a threat to one's current identifications and belief system. [13] As we have seen, in Buddhism there is a great familiarity with these higher dimensions of consciousness and with experiences and awareness in subtle and very subtle sense. Chapters 4 and 7 have presented some pointers.

A Buddhist view: The Four Maras

Now, how about a Buddhist view on defense maneuvers? Does Buddhist psychology include a kind of perspective of levels of defenses, coinciding with human development? What I find an interesting Buddhist view on the ways in which we try to escape the reality of the human condition and seeing who we are, is through the Four Maras. As the story goes, the forces of Mara (symbolizing the passions, and also used to refer to the embodiment of death) attacked Gautama, the future Buddha, on the night of his awakening. They shot arrows and threw swords at him. It didn't work their way, as these weapons turned into flowers. Canadian Tibetan Buddhist teacher Pema Chödrön sees them as ways in which we, just like the Buddha, feel like being "attacked" everyday. [14] Feeling under attack, or feeling a losing of familiar ground, we turn away and do not want to be with what presents itself here-and-now. The four styles in this "turning away," accompanying selective distortion, may be viewed in the ways that we react defensively. They may be recognized as they happen in inner mental life in general and certainly also within the meditation session. I won't give the examples from Tibetan texts and diary; by now, I won't need to convince you that they can easily be fitted in.

(1) Devaputra Mara

Here, attraction and greed lead us to "approaching" or wanting to feel good, and aversion leads to "avoiding" or inner hiding. It's about identifying with the passions. As we don't like feeling anxious or angry or desperate, we grasp for something pleasant or distracting and become addicted to avoiding pain, and attached to the defensive personality structure that this tendency reinforces. This Mara can be referred to, briefly, as: approach-avoidance.

(2) Skandha Mara

This Mara refers to a style we may use in a situation that we feel on unfamiliar ground, or where ego-boundaries are loosening. We may show a tendency to recreate ourselves and try to be who we think or thought we are. We yearn to be "the old self" and to return to the solid ground of a familiar ego-personality as quickly as possible. I like Trungpa Rinpoche's description of this: as "nostalgia for samsara." The term *skandha* relates to the "five *skandhas*" (Sanskrit, Pali *khandhas*), or "five aggregates," a Buddhist model of personality functioning. The *skandhas* show how defensive conditioning forms and shapes the ego-personality (the construct), and separate-self sense (the felt-sense). The ego-personality is continually being constructed, deconstructed, and reconstructed. The five aggregates are: 1. form, matter, 2. feeling tone, 3. discernment, recognition, 4. mental formations, volitional activities, and 5. conscious awareness. Brief reference to this Mara can be: the old self.

(3) Klesha Mara

This is the name for the maneuver in which, with our emotions, we keep ourselves asleep, un-conscious, and ignorant, with a mind full of dust. *Kleshas* are mental states that cloud the mind and manifest in unwholesome actions. In this way strong emotions, for instance anger, weave our thoughts into a story line, giving rise to bigger emotions. The narrative gets inflamed and hot by elaboration on the pain. With exaggeration and reproach, others are taken into the heat. What begins as an open space may become an inner forest fire, an outer world war. For brief reference to this Mara: emotional heat.

(4) Yama Mara

This one about life-death anxiety is the last of the four. *Yama* refers to death. All Maras in fact arise from our fear of death and the *yama*-variety is particularly rooted in this anxiety. We may wish to perfect our inner lives and feel good about ourselves as we gradually find our way and think, complacently: "that's me." But from an awakened point of view this is a state of non-developing and non-discovery, that equals death. *Yama* Mara may also be seen as connected with the anxiety to live in impermanence and imperfection, leading to controlling and closing off instead of turning to the inner obstacles of not tolerating the sense of powerlessness, and dealing with uncertainties. For brief reference: life-death.

Any recognition with one of these four?

8.4 Musings: Western-psychological and Buddhist ways in self-limitation

In these selected examples, both the Tibetan psychological Six Realms and Four Maras, on the one hand, and the Western DSM and defense approach, on the other hand, may address a range of dynamics in humans that can show up in an on-going moment-to-moment way.

While the Six Realms, as we saw, can have a more fluid, momentary sense, and a more dimensional flavor, it also has a more personal taste: it touches more a "first person, second person" feel: it may be me, or you, who is addressed. Categorizing views, like the Western DSM view, seem to be more distant, with a tendency to "crystallize" in what is seen as more rigid personality problems and disorders.

It seems that in Western ways of describing, the "resolving" of inner conflicts by way of (unconscious) defenses is taken for granted ("it's just the way we do this, this happens to us"). Yet, it can't be missed that these defenses support selectivity and distortion in perception and giving meaning to what reality is presenting to us; they support us in avoiding certain aspects of reality. The Maras' way of classifying mechanisms, and the way of the defenses, show common ground. Most of the 31 DSM defenses can fit, in some sense, in the four Mara categories. For instance, denial can be found under "approach-avoidance" (*Devaputra* Mara), and acting out under "emotional heat" (*Klesha* Mara). Defenses can also be subsumed under the roots of suffering that translate in the Six Realms: like denial under "aversion" and intellectualization under "ignorance." Distortion of reality, both by defenses and by Maras, is recognized on both sides. Yet, and this is crucial: while in the Western view of defenses the need to defend an ego-personality is taken for granted, in Buddhism the four Mara maneuvers are also seen-through in their attempt to keep up this illusion of a separate-self sense. Next to being descriptive, like the levels of defense, an important aspect in the Mara-view is that it is seen as pointing the way to being completely awake and alive. You might say: by letting go and letting ourselves die, each moment, at the end of every out-breath. The patterns described defend against a reality that is unconsciously imagined as unbearable because it signals a danger to ego-personality's existence; indeed, it may mean the end, the death, of the personality. At the same time, in the Buddhist view, these patterns are obstructing our potential to experience reality and to awaken.

Interestingly, the Western defenses approach is a scaled one, with seven levels or stages that run from high to low adaptation. Western psychology has an impressive tradition in developmental psychology. There are intricate theories, and more and more empirical research is done, regarding these levels, about human psychological development from infant to old age. As we saw, while the defenses go by stages, the Maras are more juxtaposed styles, placed next to each other. The Tibetan Buddhist approach, while using the practical *skandha* model of personality functioning, lacks a developmental psychological view.

As we saw in 8.3, defensive behavior patterns, to the extent that they have been learned, can be unlearned, says McCullough Vaillant. Might it be possible to unlearn unwholesome sense of self conditionings, all the way? For one conclusion, regarding development: wouldn't it be beautiful when accepted Western developmental views would expand into higher, more subtle ranges, with lessening self-attachment? At this place we have looked at the defenses; yet, so much inner space could be opening in further going health development and maturation!

Connecting with Settling the mind practice

Let us now turn to the relationship of personality styles, of Maras and defenses, with the practice of Settling the mind in its natural state. Certainly, all Maras and defenses have shown up, in the diaries and texts, as you could see in the illustrating examples.

The core instructions for the practice are non-distraction and non-grasping: not going with some thought that takes away presence with what is, not identifying with what comes up, and in that way let presence crumble. In my view, all defenses and Maras include both distraction and grasping. They are all, in their various ways, distorting perception of reality. Yet, Settling the mind in its natural state invites you to observe, to just be with mental space and what arises in it, in the moment, and all-inclusive. So: no delay, no preference, no hiding; just not the phenomena that defenses and Maras have specialized in and brought to great heights. While the set up of Settling the mind practice applies a temporarily chosen selectivity, for sharper training, the attitude that is practiced will generalize; more about that later.

"Being with what is," in shamatha starting with: observing what is, seems to be so difficult. It coincides with an immense un-learning and deconditioning operation that can be described from all the four Integral perspectives. What makes it so difficult, so challenging? From the first perspective, Experiences: anyhow, it requires a lot of motivation and practice. Yet ... what motivates is a gradual, subtle confirmation in a sense of greater ease, and relaxation, with greater presence and authenticity. From the Behaviors perspective: this coincides with gradual changes in neuronal networks and association fields, all-over the body-brain, neuron by neuron, molecule by molecule – which takes time. And from Cultures and Systems: "being with what is" will require a huge shift in collective consciousness. It will require a melting of collective reactivity and self-limitation. There seems as yet little collective "holding" and support for "being with what is": really being with climate change, with the deeper causes of perpetuating suffering, wars, weapon trade, hunger in the world, to name a few. No support for "seeing through," seeing how many forms of man-made suffering can be ended. Most of the time, our defenses and distortions are taken for habitual. Our collective cultural and political structures are based upon this state of affairs.

In a larger societal context

The overall approach to mental suffering in the West has mainly been, next to institutionalized religion, in terms of psychiatry and psychotherapy. Regarding neurotic suffering, an overall message in Freud's *Civilization and its Discontents* may be read in a way, that psychotherapy can alleviate neurotic suffering, but does not address people's habitual sense of unhappiness. [15] This existential suffering is what Buddhism addresses: Buddhism exposes the root of habitual, ordinary unhappiness, and shows how we may overcome it.

Western psychology, for a long time, has focused on psychopathology, and in this field contributed much to our understanding. It has great contributions in fields including developmental research, defense and resistance, cognition and conditioning, and therapeutic intersubjective dynamics, to name a few. Next to a few psychotherapy approaches, only recently there has begun to emerge a broader and clearer view that includes training the positive. Positive Psychology creates bridges. Newer integrative perspectives are widening the scope. Yes, positive and healthy states can be trained and cultivated! And this can show in measurements, and be visualized in brain scans: neuroplasticity at work! In beneficial ways, the prevention of illness and the enhancement of health are getting more emphasis.

In Buddhism it has been understood that methods for removing the negative can be of great use. However, they cannot replace methods for cultivating and supporting the positive. Best are complementary integrative approaches that address their natural synergy.

8.5 Development and discovery, states and stages

As we saw, the defenses sequence illustrates a developmental view, coinciding with stages in maturation. In this view the defenses deal with, mostly, coarse emotional states like anger, fear, and shame. Maturation means: that we use less of the immature, and more of the mature defenses. The Maras, as well, are dealing with coarse emotional states. Yet, the Maras have a place in a context of recognizing also more subtle and very subtle states, as these are included in a larger picture.

In Chapter 1.6, in the context of a *fourth turning* of the Wheel of Dharma, for modern times, the theme of meditation and psychology is mentioned. The need for integration of various manifestations of wisdom is emphasized, the need, for every person on the path, for *integration of psychological maturation and realization*. This is also, you might say, about stages of maturation and states of realization. How to better understand the relationship between states and stages?

Stages and states, connecting with development and discovery ... the last two notions came up during retreat with Alan Wallace, during the questions and answers sessions (Chapter 2.5). Regarding *development*: in our shamatha attention training we cultivate stability and vividness. One might say: we train ourselves in attaining increasingly refined states of stability of attention, again and again. We practice non-distraction and non-grasping, again and again. Trained states tend to unfold in a sequential way, they tend to follow the natural order of coarse to subtle to very subtle. We saw how the process has been described along stages on the Elephant Path, where indeed stability and vividness both increase: Directed attention, to Continuous attention, to Resurgent attention, to Close attention, to Tamed attention, and further. Trained states can be seen to present in *state-stages* – as they are named in the Integral approach. States are more momentary, stages can be seen as more enduring manifestations.

Wallace also hinted at the *discovery* aspect in meditation, that manifests specifically when we have come to more refined states, more refined state-stages in mind and awareness practices: this is about the unveiling, and the "discovering" of the innate stillness and luminosity in the very nature of awareness itself. In the more advanced state-stages of training, chances increase that we may experience increasingly subtle states of consciousness.

So, these remarks have addressed states of awareness, connected with discovery, and state-stages connected with development, in the context of meditation. Next to state-stages, also "structure-stages" in development and maturation are distinguished, in a more general way. Examples include stages in the developing capacities for a child of cognition, emotions, and morality. As to cognition and reading: a very young child cannot read. After a few years it can read a few words, and later the child reads a book. We can see stages in these abilities that coincide with structural maturation. These structure-stages can be described in terms of stages on a developmental line. We may look at development happening through all the named four perspectives. For the Experiences perspective of self and consciousness, a basic developmental line of "self" development can be mapped, with levels of increasing maturity: instinctual to magic ... to mythical ... to holistic, to integral. We may also address an individual "values line," a line representing the (shifting) values that a person cherishes: a person can develop from *egocentric (me) to ethnocentric (my group) to socio-centric (my country) to world-centric (all of us) and on to planet-centric (all beings) to finally cosmo-centric (all of reality)* related values. These levels on the values line stand for increasing care, compassion and commitment to a wider circle of beings. This is a movement of increasing maturity that we support with cultivating the Four Qualities of the Heart meditations: wishing happiness to all, in expanding circles, geographically, including those we like and don't like, including all sentient beings.

In 8.3 we briefly referred to the higher reaches in Maslow's view of human needs. Maslow describes needs as maturing *from physiological needs (food, warmth), to safety, to belongingness to self-esteem to self-actualization to self-transcendence.* For the Cultures and worldview perspective, Jean Gebser presents a developmental line from *archaic to magic to mythical to rational to pluralistic to integral* views. For the brain and organism Behaviors quadrant we may refer to levels of increasing physical complexity achieved by evolution, as: *organic states to limbic systems, to neocortex*.

The elements in this brief information on the Integral model are not further clarified in detail ... this is just meant to give some broad impression of assumed dynamics.[16] Why all this? I find these views clarifying, and feel a need to present these aspects to be able to share some additional notions with you. They relate to the themes of "peak experiences and stable recognition," and the theme of "states into traits."

Peak experiences

Regarding peak experiences: an important point of distinction between stages and states needs mentioning: while stages need to develop, or be trained, states of consciousness are potentially directly available to awareness. What does that mean?

I can experience *states* at every moment, in principle. Basically, my state of consciousness can be, for instance: awake, or dreaming, or deep sleep. My feeling-state, emotional-state, as became clear in Chapter 5, can have quite a range. It can shift from coarse anger to subtle joy and gratitude. Coarse states, both emotional and attentional, are our daily stuff. Coarse, subtle, and very subtle states of awareness have been referred to, and also reference has been made to the nondual ground of all awareness. Subtle states are potentially always accessible, as they refer to what has been all along.

Interestingly, for more lasting *stage* development, no steps can be skipped. There can't be a sudden jumping from Directed attention to Tamed attention, on the Elephant Path, neither can there be a culture with a worldview jumping from archaic to integral! The steps in-between, all rungs in the ladder need to be covered.

States can be accessed, in principle: there can be a peak experience into a higher state, like very subtle awareness. Yet, the brief access to such a state is not enough for a more lasting access to these states: one needs a lot of training. Buddhist and other meditation traditions have brought this kind of training and the more lasting accessibility of these higher states to an immense refinement.

States into traits

For more lasting effects, a shift from state into trait is required. In micro dynamics, we may describe the process as follows. As it is explained in Integral speak: "the subject of one stage becomes the object to the subject at the next stage." This may be said in the context of both meditation and of psychological development. For example: there may be moments that suddenly I can "see something about myself" in a self-reflective way that only later becomes an integral part to my self-understanding. These momentary states can be felt as a deeper knowing. In terms of the subject of the one stage to the next: in stage one I'm identified with my desire-state for that raisin. In the next, the second stage, I'm aware of myself as this desiring person, but not identified with her: there is more maturity and freedom. The formerly desiring subject now is the object of a new, slightly more mature subject that has more options for further experiencing and expression. The moment of seeing is a state-experience that only later becomes a stage trait. Many times disidentifying strengthens my familiarity with non-identification, leading to increased maturity and inner freedom.

The states into traits shift, on a larger scale of meditation research, relates to the question: when we have attained greater emotional and attentional stability during retreat, will it last? Has the frequent state-access "translated" into state-stages, "did I progress?" And then: not only in the measurements directly after retreat, but also at home? And: does this possibly increased access to more refined states influence my emotion regulation abilities? This all connects with a large research question in the Shamatha Project: do the changes (like: improvements in attention, maintaining resilience in the face of stress) persist after meditation trainees return from the retreat experience to the cacophony of everyday life? We had some answers to these questions, in fact, with the outcome data from the Shamatha Project research as presented in Chapter 1. The state into trait phenomenon also connects with my overlapping smaller scale personal research, with the hypothesis that the trained states translate into lasting traits.

So, these are many terms and notions ... The relevant thing is that I hopefully succeed in transferring to you some overall felt sense of what states and stages stand for, and how they relate, as we will be back with those in the direct context of meditation and science, in Chapters 10 and 11.

Chapter 9

Psychological turbulence and self-healing

> *While Alan's instructions for breathing and Awareness of awareness practices tell me to let go of thoughts as soon as they arise, with Settling the mind I am allowed to think!* (Diary, 2007, 9.4)

In this chapter, we will be back at the inspiration, transpiration, and the blood-sweat-and-tears of direct meditation experience. As became clear: doing shamatha practice and particularly Settling the mind in its natural state may bring up quite a few surprises. With relaxing and opening up, old stuff resurfaces from the recesses of deep mind caves. I trained in balancing the greater openness and relaxation, while at the same time observing the automatic tendency to grasp, react and contract, when things became too unfamiliar. Here I experienced what I described as a "yo-yoing process," and sometimes a "whiplash" of grasping to familiar self (Chapter 5.4).

At the end of Chapter 5, I stated that I feel connected with clients who gave me their trust in sharing their paths, from whom I learnt a lot that will also transpire in what follows. In this chapter, the theme of (unconscious) reflex defensive mental operations aiming at self-protection will continue to have our attention, now in a "lived" embodied meditative sense. Here we will get to see what shamatha meditation and specifically Settling the mind in its natural state brings in as an antidote, and an option for change, leading to more degrees of freedom in our being. Views from Western psychology and psychiatry, and Buddhist notions will give some context.

Let me start with a surprising uninvited turbulence experience I went through in the last phase of the project retreat.

9.1 Variation on a theme: body-mind turbulence, falling on my head

What follows here is a specific diary-example that I wrote on body-mind turbulences, catalyzed by falling on my head. End November 2007, with -14 degrees C, being 6.8 F: on a walk over a frozen path, I slipped and the back of my head hit the ice. A co-participant *bodhisattva* dredged me up. A flash of feeling: oh, it's all over, participation story finished. I was told that some of the final attention measurements, planned for the next day, would not be done as they could not be included in the research data since it would be influenced too much by the fall. Then Clifford Saron, our Scientific Director, asked me to write a report about what this falling did in my meditations ... which in a way felt very consoling: "this may be included in our common endeavor, this too, all-inclusive." The falling, with a brief period of unconsciousness, has been a shock, with pain, contraction, and crying. There was a brief amnesia; I couldn't remember what just went before. It brought back memories of an earlier concussion trauma. At first there was a sense of unrealness, then there was "seeing," settling, some relaxation. As it shows, falling on my head, as disagreeable as it was, has become a valuable practice for me of "taking adversity as the path," thanks to Cliff. Here are some quotes from the report:

> Day 1 and 2: I'm aware that the movement of discerning mindfulness is a bit bumpy, more than before. There's a sense of less pulses of ascertaining awareness, more holes than before of non-ascertaining ... on these possible 600 pulses per second, that Alan told us about. With less continuity and stability in attention, and specifically: less density and vividness. Still, the greatest relief, joy, of this first day is the feeling that the "gracious host"-mind has remained unperturbed, and undamaged, with all these unruly, unexpected, sometimes weirdo-guests.
>
> (After one day) I feel I can relax the voluntary muscles again (thanks to repeated monitoring "inner biofeedback" practice

yesterday). Yet, it is also clear to me that involuntary muscles are still contracted, and that muscle tonus is still increased.

One of my first checks is: do I still have the ability to get beyond coarse excitation? So glad, I can still continue staying with the breath, without losing the thread. Still, there is less completeness, and initially I'm less able to distinguish foreground and background in Settling the mind practice, there's more flatness.

Day 1, the feeling is that this is what saves me: just being present, and finding myself yearning for the space of mind, paradoxically almost grasping to the space of mind. "No center, no periphery": not only no center, also no center with a head to fall on! With a little corner of meta-cognition, monitoring, discerning mindfulness that stays clean, free, momentary. Later this day I intentionally put an imaginary button for distraction and grasping a bit wider than otherwise: I stay a bit longer with a thought, to really be with it, while monitoring; still in a way that I don't lose the space! Also for some stretches I give myself the permission to just rest in the space of mind; which later shifts more easily than before to Awareness of awareness practice. This right from the start helps me in "keeping spaces open," and I feel very grateful for it. The beauty and elegance of all-inclusiveness, in a sense of "being space": nothing can be harmed, be added, get lost.

There's the feeling that "everything" goes somewhat slower ... what? Less velocity through the neural tissues? Edema? Generating a voluntary thought is slower; while involuntary thoughts go quickly and uninhibited (maybe disinhibited).

Interestingly, I find that in the supine position another register of mental activities seems to be opened now, in comparison to sitting practice. The feeling of vulnerability combined with supine very much motivates to see the practices as important and valuable preparations for the dying process, as Alan has mentioned them to be. This may also be nourished by practicing Settling the mind-meditation in *shavasana* corpse position (a specific supine yoga position), with a vacuous gaze to the ceiling, mouth slightly open.

Loving kindness and compassion: they are very present, all these days, with great gratitude, and, to my surprise, a sort of lightness with "as it is." Loving Kindness to all those who help and are of support, and also to myself in sharing my vulnerability and helplessness, "living it." Compassion, specifically: to all those who are suffering and feel helpless with brain conditions, or are worried, have some fear in that sense (including me, a bit), and widening to all.

With some ideations about what's happening on the brain level (the Behaviors quadrant):

In the Settling practice, images come up of tearing axons and dendrites, bruising, edema … and resorption after some days. I suppose the temporary numbness has to do with that. "Pruning" and (accelerated!) "sprouting" (Explanation: pruning, referring to a loss of synapses, resulting in fewer connections between neurons in the brain; with sprouting, new connections are made). The adagium: "If you don't use it, you lose it" comes up: how wonderful to have these shamatha practices. It feels like the neural substrate-correlates get temporarily dysfunctional as receivers for consciousness information, by sudden disconnection, by reflex massive subtle contraction of micro-muscles, and edema of various tissues. The recovery, self-healing process lies in gradual relaxation and metabolizing of the dis-balance and its effects. Thank you, neuroplasticity and compensatory pathways!

The whole adventure feels to me like a recovery from a massive grasping-reflex, that has its roots in survival reflex. On top of this most primal physical survival contraction reflex, there seems to be grasping reactivity to a perceived "losing control," reactivity also to a sense of "losing (illusionary) self," and – maybe – a mental space contraction reflex, coinciding with "losing presence." Relaxation, finding and keeping balance and openness ...

I'm so happy to continue feeling familiar with all spaces! Regarding awareness, mind, body, it feels like:

The Awareness of awareness is there immediately – always already – I can stay attuned.

Access to the (momentarily contracted) space of the mind is briefly diminished, but with the help of the tools of the practices I can get back there in a familiar way within minutes, and there can be de-contracting, with awareness of foreground and background returning.

The space of body and breathing: this has been the one that most slowly recovered, inertia ... (this is also the space that is in an evolutionary sense the most "old," related to survival-reflex, from the reptilian brain; it just does its job to the best of its abilities, according to its knowingness-ignorance). In my felt sense it took me one day for recovery of voluntary muscle relaxation, two days for involuntary, and maybe three to five days for tonus-relaxation again (51).

Looking back to this piece of diary: I have been so aware that turbulence experiences always challenge a sense of self. In that way it urges for insight practice regarding self: I have felt this falling on my head as motivating for vipashyana! Probably I did well in going quite some time in supine, in my recovery from body-mind turbulence. The Four Qualities of the Heart have been flourishing.

Right after the shock I've been in some not-so-natural altered state, in a rather coarse register. I could recognize defenses like denial (this can't have happened) and a feeling "it's all over now," generalizing, exaggerating. I could sense the *Skandha* Mara of yearning to be "the old self" (Chapter 8.3). The original head trauma, going with a traffic accident that I had years before and that this fall reminded me of, has had a somewhat longer period of unconsciousness. At the same time as it has resulted in a troubled feeling, I've been aware that it also in some sense has brought a certain lightness, less attachment, more openness in my life. Years later I asked Alan about what may be the effects of a cut in consciousness contact. As I understood from him, a sudden anomalous state, with un-consciousness and "collapsing of the mind in substrate consciousness," may bring in a brief contact with substrate qualities of bliss, luminosity and nonconceptuality. Here, I understand that head trauma may be a catalyst for more coarse *and* more subtle state experiences, in comparison to the habitual degree of coarseness.

Regarding what happens with the mind-brain, some more musings come up. With the Settling the mind practice, like what was said about breath practice, a natural feedback mechanism is at work, in the mind, and correlated brain sense: observing, without grasping, leads to a thinning out of mental content. The "if you don't use it, you lose it" addresses both mind and brain, regarding unwholesome grasping habits. Non-grasping correlates with wholesome pruning, giving space for more wholesome neuronal associations.

My amateur-wonders go to a traumatized mind-brain ... might it be possible to say that, for the *state* aspect (the state experiences that in principle, potentially, are always available, unless there is very heavy brain damage) more emphasis will be on synchronization, facilitation of existing networks? And that for trained states and *state-stages*, emphasis will be on the need for forming new neuronal association networks, sprouting, cells "firing together, wiring together," with pruning of old overdue connections, cleaning up?

9.2 Settling the mind and "falling out of habitual conditionings"

And still ... also in this turbulent episode, like I described in the diaries: there is this deep sense of stillness – a lively, dynamic stillness. As mentioned in (33): underlying all, in some way surprisingly, in other ways not at all: this deep sense of joyful stillness. Which brings us to a Settling the mind practice, with special attention for this aspect: the union of stillness and motion.

This union-realization has been well described by Panchen Lozang Chökyi Gyaltsen, the tutor of the Fifth Dalai Lama of Tibet, who lived in the sixteenth to the seventeenth centuries. Alan Wallace addresses his instructions: they include the realization that while practicing Settling the mind, without suppressing thoughts, you can recognize what they emerge from and what they dissolve into. You can observe their nature. As Panchen Lozang Chökyi Gyaltsen asserts, each time you observe the nature of any thoughts that arise, they will vanish by themselves, following which a vacuity appears. In the same way, if you examine the mind when there is no movement, you will see an unobscured, clear and vivid vacuity. There is no difference between the former and latter states. Meditators call this "the union of stillness and motion."[1]

Mind: the union of stillness and motion – guided meditation

Let us do a Settling the mind practice that regards this union. This guidance has been inspired by some themes Alan Wallace addressed during retreat and also in his book with the telling title *Mind in the balance*. In this book he explores the relationship between Christian and Buddhist meditative practices. In the section where he describes ways of observing the mind in both traditions, Wallace presents this Buddhist approach, about observing stillness and motion. Regarding motion, he describes how " … most bizarre thoughts and desires may lurch up and take you by surprise." After a start, with settling the body and the breath, we proceed to Settling the mind practice. Have your eyes open. Familiarize yourself, again, with observing mental events, and with observing the space of mind and anything that arises within it. This mental space, not located inside or outside, not having a center, a periphery, a size, or borders ...

In the beginning of the practice, when you start observing

thoughts and images, you may find that they disappear as soon as you notice them. Have patience, and relax more deeply. You are on the way to discovering a place of stillness within the motion of your mind. We will open the Pandora's Box of our minds in this practice. We focus our full attention on whatever emerges from that inner space. Again and again, when thoughts arise, you may immediately be swept up by them. Your attention will be carried away to the referents of those thoughts, you will be distracted. For instance, if a memory of talking with a friend this morning arises, your attention may be focused on this friend and the circumstances involved. In that case you shift into daydreaming.

While the mind is constantly in motion, in the midst of the movements of thoughts and images there is a still space of awareness in which you can rest in the present moment, without being jerked around through space and time by the contents of your mind. This is the union of stillness and motion. The invitation is to let your awareness be as neutral as space and as bright as a well polished mirror. In that, you are observing the face of your own mind, with all its blemishes, scars, and beauty marks. This practice involves a direct path to self-knowledge.

When you get distracted and carried away by the contents of your mind, it may be as if the space of your awareness collapses to the size of your thoughts and memories. As soon as you note that you are distracted: relax, loosen up your body and mind and release your grasp on the thoughts that have captured your attention ... You can release the effort of identifying with them. You may connect this with the exhalation, a natural occasion for relaxing and releasing.

(*There are phases of stillness, and then motion again ... Wallace continues*) When you practice Settling the mind in its natural state, gradually the quantity of thoughts and images will subside. There will be moments that you may not notice any contents at all. In that case, closely observe the background of the empty

> space in which thoughts and images appear. Note whether it is a sheer nothingness or has any characteristic of its own. Doing so, you may begin to detect very subtle mental events that had previously escaped your attention. They are so subtle that they can slip under the radar of ordinary consciousness. You may become aware of mental processes that had previously been locked in your subconscious.
>
> Now, you have set out on the expedition of exploring the hidden recesses of your mind. Long-forgotten memories will emerge out of the blue, funny fantasies may haunt you, and the most bizarre thoughts and desires my lurch up and take you by surprise. Whatever thoughts, memories and images arise, simply be aware of them, recognizing that they are only appearances to the mind. Observe them without being sucked into them. These thoughts and images have no power of their own to harm you or to help you. They are like reflections in a mirror. They are as insubstantial as mirages and rainbows. Yet they have their own reality, as they causally interact among themselves and with your body. In this way you may discover the luminous, still space of awareness in which the movements of the mind occur. With that you will begin to discover an inner freedom and place of rest even when the storms of turbulent emotions and desires sweep through this inner domain.[2]

The words: " ... being jerked around through space and time by the contents of your mind," remind me of the line by Nyoshul Khenpo Rinpoche: " ... beaten helplessly by karma and neurotic thought." More words that touch me: " ... in the midst of the movements of thoughts and images there is a still space of awareness in which you can rest in the present moment." Our focus and interest is now with the mind, shifting to the mind perspective. We get to see the union of stillness and movement, of mind perspective and event perspective. We are observing the face of our own mind, including turbulences, blemishes, scars, and beauty marks: the practice offers a direct path to insight in the psyche and a more spacious self-knowledge.

9.3 The energetic dynamics of selfhood

In the Tibetan context, most strong emotional upheavals and psychiatric problems are understood as disturbances of wind or air. In earlier explorations in Western and Buddhist meaning making, the theme of energy dynamics has been touched upon. Mention was made of the "inner winds" (Tibetan *rlung,* Sanskrit *prana*), also translated as psychic energy, life-force (Chapter 4.1). Anne Klein notes that in Tibetan one term for extreme anxiety is "life wind," (Tibetan *srog rlung*), also referred to as *sok-lung*. Terry Clifford, author on Tibetan Buddhist medicine and psychology, mentions accordingly that intense neurotic behavior and the psychological and physiological symptoms of nervousness are called the disease *sok-lung,* a disorder of life-wind. In Klein's view the energetic dynamic of selfhood can be seen as a middle ground because it encompasses both the conventional, empirical self, as we are familiar with in the Western paradigm, and the ultimate self. While in Buddhism the psychological self is affirmed at a conventional level, a fixed, permanent self at the ontological level is rejected. [3] This brings up the question: in what ways might the perspective of energy dynamics support in understanding my experiences, including mental turbulences, in their personal and transpersonal manifestations? And: how do the energetic correlates, felt-sense, relate to "signs of progress" in the practice, and "signals for worry" in the sense that with certain *nyam* I may rather seek a doctor? While not really having been worried, I have experienced the feel of both these signs and signals.

The felt sense of opening and closing

Let me reflect on the questions raised, taking my own experiences as addressed in Chapter 5 and Chapter 9.1 as the starting point. While in making sense of these experiences I can't avoid drawing some meaning from my background in psychology, my remarks at this place do not address Western diagnosis and categorization.

The direction and feel, for me, of a "falling out of habitually conditioned personality" have often started with increasing relaxation. The relaxation has invited for opening up into unfamiliar ranges of experience. This, in some instances could arouse anxiety, and a variety of (reactive) *nyam*. With that, the outcome in energy dynamics could be, mainly, in the direction of movement 1. or movement 2.

1. On-going relaxation, opening:

I describe some experiences in Chapter 5: a more gradual process, with *yo-yoing dynamic*, with micro-contractions mindfully embraced within on-going relaxation, trust, like described in (24): spikes of moments of anxiety arising-being and further melting-dissolving, first with somewhat larger spikings, then getting smaller, more dissolving, more and more relaxing. This goes with decreasing grasping, with a sense of clarity, joy, cognizance, (41): openness, glimpses in knowing things to their nature, now, without conceptual overlays. Trust. In other cases, this could be mainly in the direction of:

2. Contraction, closure:

This might just involve the micro-contractions as part of the yo-yoing spikings. However, when the amplitude and associated sense of threat was too big, there was an intensifying reactive grasping reflex, up to a *contractive whiplash*. Like in (46): there is a "finding myself back" in a subtly tense, contracted state in the body-mind. The space of mind might contract to small, as if forming a membrane (47). As I described, a momentary paranoid re-structuring of the personality came up as a *nyam*; gluing self fragments together again, with anger (30). As to the body, I'm reminded of the Startle Test research we did for the measurements in the project (described in Chapter 1.4), involving a reactive pulling together of the body, with muscle contraction. There is contraction and limitation.

Signals for worry? Manifestations, connected with both

movements 1. and 2., in various persons, may include: confusion, anxieties, agitation, body pains, and passing paranoid ideation. Can we remain with micro-contractions as part of the yo-yoing process, in ongoing opening? For me, they have felt like passing momentary graspings that could melt right away. This is the melting, or deconditioning, of a grasping reflex.

The falling out of habitual conditionings, with a dissolution of the personality matrix, to my feeling, always leads to a transformation, and it always has a potential for greater health. In the process, a unique momentary dynamic mix of micro closure and opening movements can be recognized. The way in which this momentary outcome, moment by moment, can be contained, represents a balance. With non-distraction and non-grasping, manifestations generally can be lived and viewed as signs of progress; with grasping and attachment, manifestations may evolve into signals for worry. The felt sense of this manifestation may differ for the one who is living it and the onlooker.

Contraction, relaxation and re-orientation

The simple diagram in Figure 3, jotted down by me during the shamatha expedition, illustrates some dynamics as described in the preceding section. The figure is based on first person experience in Settling the mind practice. It shows one possible correlation out of many, in one possible model. It also illustrates what I addressed in the video interview, in (49).

Here is some clarification to the figure, mostly in third person language, about first person experience; making the figure more generally accessible. This is a diagram on just one axis, the vertical axis, pointing to measures of *contraction* and *relaxation* regarding body, mind, heart and awareness. Measures of contraction, disbalance, ignorance and exclusion go together ("up" in the diagram). Relaxation, balance, knowingness and inclusion connect with "down" in the diagram. On the side of contraction, there is a relation with cramped identification, over-identification, feeling like one

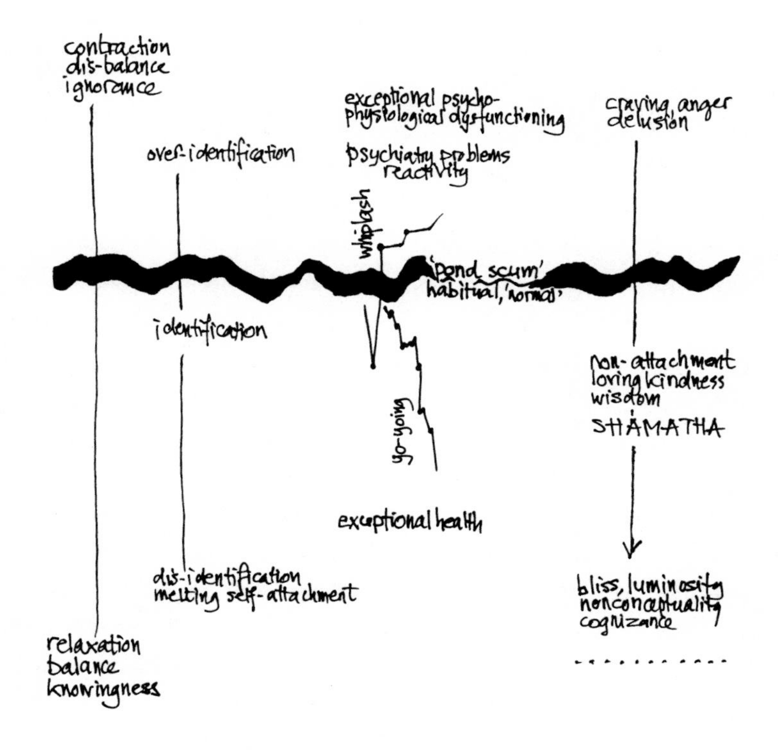

Figure 3: Diagram regarding physical and mental functioning, correlating with energy dynamics

bunch of grasping, with a felt sense of hardening, crystallizing, freezing. On the side of relaxation: an openness with non-grasping for identity, non-identification and dis-identification. There is a sense of melting of self-attachment, of fluidity. The irregular thick horizontal line of "pond scum" (described in Chapter 5.3, nr 16) correlates with the limited band-width of consciousness, that is conventionally seen as normal, connected with the habitual unquestioned sense of a separate-self (the expression "pond scum" for this I heard from Alan Wallace). Possible states can be seen to correlate with different vantage points.

When falling out of habitual conditioning there is a brief moment of "chasm": no hold, no dust, no ignorance; there is openness to truth, to wisdom. There is freedom, this brief moment is prior to the arising of the self-contraction. Falling out of habitual conditioning may be experienced as a "dark night," a dark night of the soul. Dark night, generally, is a reference to a passing through, or a letting go of attachment or addiction to a particular realm or state of consciousness. When some self sense returns, there may be the described micro-events of *yo-yoing*. When yo-yoing is embraced in relaxation, we see a spiking line going down, with some micro-reactivity, and further relaxation ... small amplitudes; movement 1. Manifestations, including *nyam*, may be seen as "signs of progress." Alternatively, and sometimes alternating, more intense reactive *whiplash* may show up: with some losing of personality matrix, a reactive grasping contraction reflex may occur: the larger mark, after some down, up, crossing the pond scum line; movement 2. [4] There is a loss of trust and a sense of threat, of being overwhelmed. The sense of threat may regard inside or outside; the inside threat may be projected to outward, like in a sense of paranoia. Transient phenomena become more fixed. The movement, in its evolving, may show in emotional, maybe psychiatric problems, up to exceptional psycho-physiological dys-functioning. There is increasing contraction, sense of separateness and duality. Manifesting experiences, *nyam*, have evolved into "signals for worry."

The zone to the right, including the word SHAMATHA, correlates with some notions from Tibetan Buddhist worldview. In a life, lived in habitual ignorance, the three aspects craving, anger, and delusion are seen as the three underlying poisons. They cause dis-balance, and with grasping they get reinforced toward increasing dis-balance. The down arrow shows one possible way out, in this context: shamatha training, attention practices and The Four Qualities of the Heart. When obscurations melt, and dust dissolves, bliss, luminosity, and nonconceptuality are experienced, ascribed to the substrate consciousness. This coincides with exceptional health, with a radically changed body-mind.

Mental rebalancing

You may wonder: do we all have to go through these turbulent *nyam*? How much might this be influenced, in Western terms, by hereditary predisposition? Or in Buddhist view by karmic factors? Do we all "have" these *nyam*, does it say something about how much I'm hiding for myself, how defensive, reactive, maybe neurotic I am? Well, for a start: as mentioned also by Lerab Lingpa and Düdjom Lingpa: *nyam* are very different for different persons. And such is the same with the rate and ways of their manifestation. Some persons just go smoothly through the process.

The way I see it: when persons go through their (strong) emotional, psychological reactivities, dynamic, momentary *nyam* are bound to come up, let's name this the *first trajectory* of more coarse entanglement-attachments. The *second trajectory,* being the more subtle, with unconscious self-attachment reactivities, often is more prominent later on the path. This may be the more tough, through pervasive cultural and individual conditioning. It brings in anxiety for "Westerners," for whom identification with and attachment to the separate-self sense is so much conditioned normality. Of course, the two trajectories do overlap and require dynamic integration, moment by moment.

In the process Settling the mind practice facilitates a deep mental re-balancing process: it includes the training in tolerating anxieties, seeing mental objects for what they are, without needing to grasp to, and identify with them. The fundamental point is that with non-grasping we create a sense of wholesome "stepping back": not by dissociation but by de-centering and witnessing, facilitating reflexivity. It is not about the turbulent thought's content, it's about our relation with the nature of thought. There can be more options than grasping and being hung up – there can be a choice, with more degrees of freedom. In the second trajectory, in my view, with nongrasping and nonidentification, some insight in the psyche is coming in: some "seeing through" the illusion of separate-self sense. We are not going to deeply investigate this in shamatha; yet, if one comes in contact with this separate-self sense as being an illusion, some may accept and comprehend. Some others will initially be in reactivity, pain and panic. Then, a person with some contextual understanding, trust, a meditation path, a good teacher and a supportive group will generally be able to face the turbulences in a non-grasping way. It is good to realize that also this grasping reflex, on its level, is meant to "protect" us – even if this works out in unwholesome ways. In my experience in various retreats, most persons, so to say persons with an average degree of inner groundedness and resilience can let old neurotic knots and self-attachments "melt" by observing and not identifying with mental content. However, a person that gets fixated in denial and repression may fall into mental problems. When contraction and closure get reinforced and intensified, they may evolve into a state of "psycho-spiritual emergency," that may manifest as what is called, descriptively, a psychotic episode. Whatever the label: these persons have – maybe ever so briefly in a sense that they may not be conscious of it afterwards – tasted that some more free state *is* possible. And also, what is then named a mental illness is a manifestation of our true nature. It is a distorted manifestation, we may say, with a lot of dust over it; yet, a manifestation.

Emergence and emergency

After these remarks that connect with felt sense and meditation, in what follows a few more objectifying considerations are presented. While the theme of spiritual emergence and psycho-spiritual emergency is vast and complex, these are just a few remarks. As we participants in the retreat also experienced, three crucial questions relating to coping with this loss of familiar self experience include: One, is there enough resilience, balance, basis with this person, with me, at this moment? Two, do I have the support of a teacher and some understanding, some map, some way of giving meaning to the experience? And three: are there people around who can offer a compassionate holding environment?

Settling the mind practice may offer an emblematic understanding of what may happen, and: what can be the power of witnessing. Signs of progress, and signals for worry: it starts with "manifestations" that catch our attention, and we don't know yet what will evolve. Here are some remarks, based in Western psychiatry views, regarding the probability of manifestations as being "signs of progress, or signals, where worry is needed." Some differentiation may be made. There is a higher risk for being signals for worry when there are psychotic episodes that include the following: 1. they are more intense than so-called normative religious or meditative experiences, 2. they are experienced as terrifying, 3. they are very preoccupying, 4. they are associated with deterioration of social skills and personal hygiene, and 5. they involve special messages from religious figures.

On the other hand, when the situation includes the following four aspects: 1 – there is so-called good pre-episode functioning: the person functioned well in his or her life, 2 – there is an acute onset of symptoms during a period of 3 months or less, 3 – there are stressful precipitants to the psychotic episode, and 4 – when the person has a positive exploratory attitude toward the experience: then chances are higher for being signs of progress. The manifestations may include ecstatic mood, a sense of newly gained knowledge, and delusions with spiritual themes (which most psychotic disorders do not include). [5] Also, the absence of conceptual disorganization and confusion can be named as a positive sign.

With shamatha practices, we have a focus, which generally gives some hold, in comparison to other, more open practices. Still, as Düdjom Lingpa addressed, also with shamatha there can be far-ranging *nyam*. And worrisome dynamics, even for shamatha practitioners, do not stop at the named shamatha phenomena. They may evolve a variety of expansive energy shifts, with further alterations in physiology and awareness. An example is what is now often broadly referred to as *kundalini*, originally connected with the Hindu tradition, understood in terms of *prana* and the chakra system.

I have been involved with a number of co-participants in retreats, who went through psychospiritual emergency. I'm reminded of a man on long term retreat. He was very motivated, sometimes sat whole nights. When you do this with a closely knit group, like sometimes happens in Zen, you "carry each other," in some sense, in the collective energy field. The teacher keeps an eye on every participant, and gives advice. This man became very agitated and anxious. Often stress, possibly underlying stress in one's life, and effort in meditation may start off graver problems. Most of the time more grounding is needed, as our psychophysical make-up is not used to long stretches of sitting still in silence, missing the habitual distractions. Grounding, movement, and "cutting carrots" so to say, helping in the kitchen, "sawing woodblocks for the fire," and explanation within a supporting relationship, help create a context of holding and meaning for the experience. A woman, on shamatha retreat, almost exclusively did Awareness of awareness meditation, without having the foundation of deep relaxation, and robust stability. Vividness may get very enhanced, but there's reason for worry when there's a lack of stability. She was spacing out and described derealization, depersonalization experiences. Then better stop for a while and cut carrots. It was a moving experience to help her ground, by giving foot baths and foot massage, and to go for walks together, making a practice of placing our feet down very aware and with force. A man, Vajrayana practitioner started seeing deities around, and behaving in ways that others couldn't follow, while neglecting to eat and take daily care of himself. Sometimes, when rest, care of the group, grounding practices and so on, are not enough, some medicine may be required for a brief period of time. This again is a balancing act: normalizing the sleep pattern (with the help of sleeping medicine) may be enough to support inner resilience and preclude deepening of anxieties and reactivity conditionings. At other times this turns out to not be enough and consultation with a psychotherapist or psychiatrist may be needed. If possible, of course, this professional has some meditation and transpersonal interest and experience.

Too often a meditator feels abandoned, when in need of structured care and support by the teacher. Too often, there is either under-reaction (not providing the right care, as needed), or overreacting (out of panic, embarking on a heavy psychiatric treatment path). Training for meditation teachers in this field, including facing one's own narcissistic vulnerabilities ("this can't happen in one of my groups!"), would be advisable. This will be for the benefit of all: the meditator-in-turmoil, the teacher, and the group .

Stan Grof, psychiatrist and one of the founders of Transpersonal Psychology, notes: "It has been hard for mental health professionals to accept the healing and transformative potential of certain mind states similar to those that are traditionally treated as pathology." [6] It is unfortunate, that currently, such experiences are almost always taken as a sign of mental illness. What might be potentially beneficial and healing experiences are then mostly responded to with some kind of ignorant suppression. These crises are better seen as a kind of non-pathological developmental crises that can have deeply transformative effects for a person's life. It is crucial that, if possible, with the right support such crises are allowed to run their course to completion.

9.4 About Settling the mind and therapeutic effects

In relation to Settling the mind practice we have considered dynamics in broad terms of relaxation and contraction, as basic movements in the body-mind. From that, with the notions of signs of progress and signals for worry, we expanded the focus to larger dynamics in transformation: with exceptional health, on the one hand, and exceptional psychophysiological dys-functioning on the other. In the former section, the potentially problematic side was addressed, including phenomena that may include dysfunctioning. Yet, in a larger context of spiritual transformation they may be viewed as presenting manifestations of a temporary non-pathological crisis. After these considerations regarding macroscopic emergence and emergency, back to the more microscopic inner dynamics in Settling the mind practice.

In the course of practicing Settling the mind, it becomes quite clear, also in felt-sense, that the habitual thinking, associating, daydreaming sort of thinking is quite different from what we may see as the bare activity of the mental sense. The last, the sixth sense in Buddhism, refers to a sort of "clean" mental cognizing and discerning, that is stripped of all grasping and identifications around it. The un-reflected grasping, chaining, associating, kind of thinking with which we generally identify reinforces our entanglements. This kind of grasping thinking can never be an instrument to solve its self-created problems connected with both the named first trajectory of more coarse entanglement-attachments, and the second trajectory of more subtle self-attachment. This kind of thinking needs to be transcended, into the clean mental cognizing.

Following are some of my diary musings, right in the beginning phase of the Shamatha Project retreat:

Diary, September 2007

> There is something peculiar about this Settling the mind practice … Starting this practice, there has been some kind of sigh of relief for a part in me that wanted to "continue thinking." While Alan's instructions for breathing and Awareness of awareness practices tell me to let go of thoughts as soon as they arise, with Settling the mind I am allowed to think! However, soon it becomes clear for this part in me, that the instruction of non-grasping to thinking takes away a lot of the habitual "thinking-fun," that for this part is in thinking with elaborating, associating.

What the increasingly clean discerning activity in Settling the mind invites, is described by Wallace as a kind of luminously clear, discerning, free association of thoughts, mental images, memories, desires, fantasies, and emotions. We are plumbing the depths of our own mind. Through the lack of suppression, phenomena that had been hidden are unmasked. He refers to Settling the mind as potentially an extraordinarily deep kind of therapy, with letting the knots of the psyche unravel themselves, as the extraordinary healing capacity of the mind reveals itself. For Wallace, this is the path to deep sanity. [7]

With Settling the mind in its natural state, we observe arisings of thoughts and emotions, including inner ways of managing the arisings, in patterns. When grasping decreases, threats to separate-self sense are unconsciously defended against: for instance, they get to be put away out of consciousness, or distorted. These hidden micro-graspings correlate with a sense of un-ease in the body-mind. The energetic felt-sense training in distinguishing grasping from non-grasping is helpful here. Being present with this felt sense leads to a melting of the defensive grip that keeps the deeper nondiscursive materials hidden, including the self-attachment grip. When formerly hidden "shadow" materials come out of hiding, we sense the manifestations in the energetic correlates. Mostly, after possible temporary un-ease there will be more relaxation and ease.

In the course of Settling the mind practice, week after week, I could sense a clear shift. Witness in wonderland! With numerous repetitions of this practice, traits evolve that support the character trait-aspect of resilience. As a general finding: by going through the turbulences, I am able to train my capacity to cope with the turbulences (do just that!), for now, and for the future!

The self-healing and re-owning of hidden "shadow" materials can be quite extensive, I believe. However, the gentle unraveling-melting self-healing may not be enough for melting and dissolving of very strong self-reinforcing defensive knots, entanglements and fixations, for instance around trauma. It may be that there are no "opening" or "closing" movements, so to say, that there is no dynamism in the process, no progress in meditation – that there is just a sense of being stuck. Tenaciously repeating grasping-patterns in this sense of stuck-ness or in *nyam* may show the way where it might be advisable to do more robust therapeutic work with a supportive "outer witness"; in that case it may be a good idea to consult a psychotherapist. In the West we are socialized with psychological contextualizing. Most people in Western countries have been brought up with some rudimentary psychological way of looking and meaning making in developmental terms. The experience is that it does help: a form of psychotherapy that offers an additional way of seeing and understanding of the origins of the need for hiding, and of the structural developmental history. It can support re-owning hidden stuff. So, in this context I think that Settling the mind practice may have its contribution, in two ways. The one is in the sense of non-grasping, and presence with arising, being, and then melting of emotional contractions. The second is in the way of showing where more psychotherapeutic work, with an outer witness, may be required.

Seeing our projections

Next to my own musings, I'm happy that here I can present some musings by another participant in the Shamatha Project. They were signed up in a book by Thea Singer, who interviewed Maura about the way her participation in the Project had worked out for her. Here, Maura describes her version of experiencing the healing capacity of the mind:

> "After the retreat, I could *see* my projections", says Maura. She became acutely aware that she – and no one else – was responsible for her feelings, regardless of how others treated her. The insight extended to parental blame. "I got really bored with that storyline and focused on my own agency," she says. "And I realized how the normal chatter in our heads actually maintains an identity for us, defining our image of who we think we are." The Shamatha Project allowed the chatter to be not silenced but hushed, clearing her perceptions. These days, Maura has greater patience and trusts her intuition more. "Before, when I didn't

meditate as much, I'd easily lose perspective, get caught up in the emotional second," she says. "It was like being buffeted by the wind - trying to control things and being thrown this way and that. Now I watch from a distance and choose what to try to change and what to walk away from."[8]

Here, Maura gives a clarifying example of "the subject of one stage becoming the object in the next," as referred to in Chapter 8.5. While previously she had been identified with the storyline, now this storyline became an object for her expanded, more mature sense of self.

Deeper layers of the psyche revealing

Here are a few remarks by various authors, relevant in the present context, regarding self-consciousness, and regarding the attitude of non-grasping.

The clearing and subtle unraveling in the second trajectory unavoidably includes a sense of loss: of losing control and of losing (illusionary) self. As for myself, in practicing Settling the mind meditation, I could be witness to the dissolving of and re-arising of self-consciousness. I described in the diary (34): more and more into a vacuum, feeling some fear, this must be anxiety of losing self. Is this *horror vacui* – terror for emptiness? Almaas, originator of the Ridhwan School and the Diamond Approach, describes this process on the crest of the wave. In his words, meditation practice may "expose the first arisings of the ego-consciousness – the background of all attitudes and positions. We become able to perceive the beginning movements of ego – the tendency to go someplace, the impulse to desire, the impulse to reject, the impulse to hope. And just the clear discernment of these impulses annihilates the ego, for by now we are very near the translucence of our true nature."[9] Beautiful wordings for a yo-yoing process in action. When the organism is not ready to bear the agitation and anxiety going with the "downwards" micro-movements in yo-yoing, reactive coarse whiplash "up" may come in. Jack Engler quotes the psychoanalyst Herbert Fingarette: "It is conflict or anxiety that turns us back to check up on and reassure ourselves, and in so doing literally brings a 'self' into being as a structure separate from its experience."[10] Our exploration of the deeper layers of the psyche reveals, according to Almaas, that "defenses are still present and are in fact employed extensively. They become more active, or rather more consciously active, in the deeper stages of inner realization, revealing, in the presence of every ego individuality, structures that are, or are *similar to, psychotic, borderline, narcissistic and schizoid structures."* As he states, the practitioner does not usually really come into trouble, when these structures emerge in consciousness, "indicating that they are not the dominant structures in the personality, but they do cause considerable distress and anxiety. Thus we see that, although in the normal individual the well-adapted or "conflict-free" segment of ego predominates, the structure actually contains all the forms of the major pathologies, both structural and neurotic."[11] Indeed, I recognize and feel recognized. Isn't that interesting: we have potential access to all states of consciousness, at any moment. The access regards not only the very subtle, but also the very coarse!

This may be a place to honor two of our old Western psychological teachers, who, in their time and context, devised ways of "plumbing the depths" of our minds. I am struck by the parallels between the ways that our Settling the mind masters have instructed us, and what has been recommended by Sigmund Freud and Carl Jung. Jung emphasized the necessity of witnessing and noninterference: "The art of letting things happen, action through inaction, letting go of oneself ... became for me the key that opens the door to the way. We must be able to let things happen in the psyche." And Freud instructed us: "Act as though ... you were a traveller sitting next to the window of a railway carriage and describing to someone inside the carriage the changing views which you see outside."[12] While there are many ways of giving meaning to what evolves: they both have recognized the power of nongrasping, its power in opening doors ...

A Tibetan view on mental turbulence and health

Connecting with earlier remarks on energy dynamics in the Tibetan Buddhist view on mental health: here are some additional elaborations. Psychological understanding in Tibetan Buddhism is surprisingly modern in its view of the dynamic nature of psychological processes, Terry Clifford notes. It recognizes no permanent self-entity, there is "just an impermanent, changing flux of reactions and perceptions, hopes and fears, habits and motivations, which we project on the external world (and in so doing, manifest it) until we recognize the dynamic process for what it is and how it arises. With that recognition we can enter the path to enlightenment." This recognition forms the basis. This, for her, is the crucial point: "The psychological basis of insanity is the same basis for enlightenment. It all depends on whether or not it is accepted and comprehended and ultimately worked with as the key to liberation. If it is not, it becomes, because the realization is still there subconsciously, the cause of denial, repression and ultimately, mental illness."[13]

How to relate to the Buddhist view that the psychological basis of insanity is the same as for enlightenment? Tibetan Buddhist teacher and psychiatrist Edward Podvoll, while confirming that the "seed of madness" is present in everyone, says: "There is another seed within us, even more important ... it is the seed of sanity, a human instinct of clarity, present in everyone as a brilliant, clear awareness capable of spontaneously cutting through the self-deception of madness," adding: " ... alongside and embedded within psychotic suffering, there exists always a potential clarity and openness of mind and heart."[14] Could Western psychiatry take this as a working hypothesis? Also the person with mental health problems has, and shows, the potential for awakening. Podvoll, by the way, after having worked for a long time as a psychiatrist, later lived for years as a Buddhist teacher at Shambhala Mountain Centre, Colorado, the place where we did the Shamatha Project retreats. He died in 2003. In many ways, I feel very grateful to, and connected with him.

This again is the question, with a new accent: how do we relate to what is. Our fears not only regard emptiness and loss of separate-self sense – felt like dying. They also regard the fullness of our being, our radiance and power, and their common source.

Diary after retreat, 2010

> The feeling: when separate-self sense can accept the opportunity (and from its ego-personality perspective, this ordeal) to melt more, and become less "dusty" ... then indeed there will be more readiness toward a serene "dying" process. Melting may progress into a serene separate-self sense death with insight practice. Of course, in seeing through separate-self sense in insight practice, nothing is dying, and nothing is lost – this is just a way of speaking. Obstructions to our fullness are dying. The words depend on the vantage point: fearing death or being the fullness ...

A fundamental re-orientation on what is health

We are in need of a fundamental re-orientation and re-calibration on what is called health, and optimal health: optimally healthy not in some range of normal habitual but with the pliancy, the suppleness that is possible. In the process of this development, evolving, and cultivation, turbulences may be expected. More understanding of the dynamics and possibilities of "holding" these turbulences is needed. Let's look with an Integral four quadrant-perspective. In relation with participation in a meditation retreat, with Experiences perspective we may think of good preparation of the body-mind, and some understanding of possible dynamics, in advance. Regarding Cultures we'll see the importance of creating a safe holding environment. As to Behaviors, it would be good to develop greater understanding of fine-tuning in doing meditation and body-practices; and, where emergence really evolves in emergency, if unavoidable, temporarily taking the right medicines for the body-brain. Regarding Systems, we may think of creating a compassionate protocol for in case of spiritual crisis, and of building into the system a safe place where a person in turmoil can rest, with compassionate support.

Anyhow, connecting here with earlier remarks, in Chapter 8.4: it is a wonderful development that Positive Psychology starts to include wider dimensions of consciousness, health and happiness to the generally accepted band width, that it takes meditation (including turbulence) and phenomena like genuine happiness, compassion, hope, authenticity, and positive ethics as respectable focus for empirical research. The title of Matthieu Ricard's book is saying: *Happiness, a guide to developing life's most important skill*. This is skill training! References in his book include research, showing that positive emotions broaden our thought and action repertoire, including joy, interest, contentment, and love. Scientists believe that developing positive thoughts and attitudes offers an evolutionary advantage, as it supports us in broadening our intellectual and affective universe, and in opening ourselves to new ideas and experiences.[15]

9.5 Summarizing special contributions of Settling the mind practice

Regarding special contributions of this mind practice, I will summarize three aspects here. First, there is the way that Settling the mind has its special contributions to the shamatha attentional practices. Second is the way the practice provides a bridging function in connecting concentration and insight practice. The third aspect regards the self-healing qualities in Settling the mind practice.

It may be said that some are more in favor of this mind-practice than others, and I can understand certain concerns: as it is a practice that includes thinking mind, there are pitfalls. Some people hold their thoughts as real more than others; for them this practice, initially, may not be well-suited. Generally, it requires quite some discipline to not be seduced into a juicy daydream, or not continue on a promising thought train. A breath-body practice may feel like more "other" and distinguishing. This said; here are some positive contributions:

Settling the mind and attention practices

While shamatha in general has its contribution toward helping us with "first things first" (first relaxation, then stability, then vividness), in my view, Settling the mind practice particularly adds the following accents:

One, there is the *broad range in progression and shifts in views*, that are consciously witnessed as to (micro) trained states and *state-stages*. The object may range from a clunky thoughts-knot, with sticking associations, repeating ruminations, that one is identified with, to a scarce gentle little thought pulse arising and right away dissolving in a serene space of mind. Stillness. As to the subject: the process involves increasing non-distraction, non-grasping, non-identification: these are the instructions that translate into what descriptively evolves. This coincides with the training in shifting focus: focus may be on event (thought, emotion), and it may be on awareness (space of mind), and it may shift. Also, the practice may be done from "events," or from a "mind" perspective. Gradually, when events are less, mind-perspective will become more prominent. Increasingly there is a merging of subject and object in awareness.

Two, the process is evolving under a *transparent bandage*, as Alan Wallace named it (Chapter 2.5): it is happening under one's eyes. This transparency, and coinciding "knowingness" of the practitioner, may facilitate abiding in possibly more subtle states of consciousness. The practitioner can be well aware of this, including of shifts in state-stages.

Three, all kind of mental processes, including *nyam,* may be perceived under a transparent bandage, thought and emotional patterns in the body-mind, that often involve a clear felt-sense in micro contraction and relaxation, in more advanced stages. Refining perception may facilitate unfinished business showing up. Often unraveling, self-healing and inner balancing processes find their way, into new integrations, with loosening separate-self sense.

Four, as contextualized in Essence practices: there seems to be *less of a "shamatha–ceiling,"* that might preclude further development and evolving. Settling the mind shows a natural transition into insight practices.

Five, in my experience: Settling the mind has a special contribution, also for *daily life off the cushion*: the increasing ability for witnessing and observing thought while not grasping and reactively acting, bears fruit in an overall way and can also be cultivated, as an attitude, off the cushion. There is greater mindfulness, equanimity and inner peace: not only as states, but as traits.

In a general way I would like to add here that, in my experience, to optimally cherish the fruits of this practice, it is good to combine it with body-breath practice. In that way the risk of being too much "in the head," with potential pitfalls is lessened.

Settling the mind, connecting concentration and insight meditation

As the Settling the mind practice has a selective focus, it nominally is a shamatha concentration-calm practice. Yet, as having a mental kind of focus, the shamatha practice of Settling the mind in its natural state nourishes aspects of "insights in the psyche" and mental operations. Vipashyana insight practice can be said to have a different agenda. While vipashyana cultivates insight in the nature of reality, in a wider existential context, it may include offering insights in the psyche.

In Settling the mind practice, in my experience, the observing of thoughts as events, and the non-grasping and non-identification, lessen self-attachment, and so does the space of mind, observed as having no center or periphery. Both familiarities facilitate the transition into insight practice regarding emptiness of self. Increasing familiarity with "appearances, arising from the substrate" facilitates experiencing, in on-going practice, their being nondual. This is a practice of familiarizing with reflective function, a training in reflexivity. We may remember the words of Wallace, that this is a practice of shamatha, designed to bring you to the substrate consciousness as a foundation for realizing emptiness and *rigpa.* Yet, as an unexpected dividend, Settling the mind may also offer some very profound insights into impermanence, suffering, and non-self, characteristic of vipashyana practice. [16] With these insights, and increasing openness in the space of mind, with less grasping and identification, there is space for surprising new information "out of the box," in a context of natural compassion.

Settling the mind, self-healing and therapeutic qualities

I want to start by quoting the words of Alan Wallace about the "therapeutic agents" in the practice that lead to this increasing sanity. According to him, three obscuring mental habits are "surgically removed" – nice expression – by the Settling the mind practice. Wallace describes them as follows:

1. Thinking that the thoughts and images you experience exist outside of your own mind.

2. Compulsively responding to those mental events with craving and aversion, as if they were intrinsically pleasant or unpleasant in and of themselves.

3. Identifying with them as if you are the independent agent who created them, while viewing them as intrinsically "yours" simply because you alone experience them.[17]

How interesting, again: Settling the mind is a concentration-calm practice, and not an insight practice. Still, with some guidance and mainly just observing: many habitual assumptions that are much ingrained in the Western worldview, come to be questioned!

Next to these fundamental limiting habits, let me add some complementary aspects, relating with resilience and "insight in the psyche." In Settling the mind practice, with observing arisings of thoughts and emotions, we get in touch with inner ways of patterning these arisings. With grasping decreasing, threats to separate-self sense are unconsciously defended against, in hidden micro-graspings. With the energetic felt-sense training we have learned to distinguish grasping from non-grasping. With presence, a melting of the defensive grip occurs, including of the self-contraction per se, leading to a deeper sense of ease and "at homeness." When patterns keep repeating tenaciously, working with a therapist may be advisable. The practice allows the uncovering, the bringing to the light, that which has been there. The turbulence is in the reactivities of self-grasping.

So, altogether: a rich practice, with elements of bridging functions in many ways ...

Chapter 10

Contemplation and science, some considerations

> *Reading about meditation in these third person terms reminds me of the different felt sense of doing the "lines-test" myself: ... first person sitting with the computer, in this dark little room.* (Diary after retreat, 2011, 10.4)

In this chapter we will look at some aspects of attention and consciousness research in Western terms. When talking about science, I refer to empirical science, meaning that the knowledge in the various scientific fields must be based on observable phenomena and capable of being tested for its validity by other researchers working under the same conditions; which covers a large field. This brings us back to the remarks made in Chapter 3.6, about "good knowledge." Three basic aspects have been named: first, "if you want to know this; do that," the experiment, which to me includes observing in a really fresh way. Second: there is the experience, and result. Third is the communal confirmation or rejection by those who are entitled to make this evaluation. Globally, in the empirical sciences two major groups are distinguished: natural sciences, which study natural phenomena (including biological life), and social sciences, studying humans and societies. Human direct experience can be counted under the studying of humans. Mostly this studying has been done "from the outside" (observing humans doing this or that), but more and more research is added with experiences "from the inside." Diary description can be a source of data. We may connect with the Four Integral perspectives as described in the Introduction. The study of humans and societies resides mainly with the perspectives of Experiences (self and consciousness, first person) and Cultures (culture and worldview, second person). These include:

studies in subjectivity and intersubjectivity. The study of natural phenomena globally connects with the Behaviors (brain and organism, third person) and Systems (social system and environment, third person) perspectives; they include studies in objectivity and interobjectivity. All four correlate, in viewing one phenomenon from various perspectives.

While in earlier chapters often the emphasis has been with the Experiences dimension, in this chapter these will be correlated with research, mainly, as residing under objective Behaviors. Some relevant examples are given of applied, clinical research, and some instances of academic investigation are presented. In the context of the last, two meditation styles are described, that have been researched with scientific rigor. A model protocol for including first person phenomenological descriptions in broader meditation research set-ups is presented, which I complete in regard to Settling the mind in its natural state. I report on two simple pilot-explorations. The references to published research, going with these considerations by an interested research-guinea pig, are in no way representative. The research named caught my attention, partly because it has been effected by persons who in some way have also been involved in the Shamatha Project.

10.1 About a map and five attentional abilities

According to the Integral approach any event can be "mapped" in four ways, as we saw. The mapping model, with some more elaborations, is a neutral framework that can be used to bring in more clarity, care and comprehensiveness to virtually any situation. It can be used by any discipline. The model can provide a language for a dialogue between the many disciplines that will be helpful in the context of many fields of science, each with their own professional terminology, language and culture. The application has been described as: to cultivate body, mind, and spirit in self, culture, and nature. The Integral approach offers a map, a framework, and a practice. While a map is a map, and not the terrain, it has helped me to once in a while look at this map, being in the process of writing this book, as mentioned in the Introduction. In this I learned a lot, specifically, from Ken Wilber's works, and from Sean Esbjörn-Hargens.

The following aspects directly regard the endeavor of seeing one phenomenon – in this case meditation – through various lenses of different (scientific) disciplines. Integral Theory understands each of these four perspectives as simultaneously arising. Every event deserves to be seen by way of all four perspectives. None of them can be separated, because they "co- arise," they are there at the same time, and inform each other. As we saw, there is meditation, experienced by me with increasing continuity in attention (Experiences). There is meditation that can be visualized in certain EEG patterns and dopamine flow (Behaviors). There is meditation, during a retreat, in a context of social support (Cultures). There is meditation, performed in a place with a cold mountain climate (Systems). All correlating, all at the same time, all showing their aspect of the truth of the moment, while none of these perspectives can be separated, or reduced to one of the others.

The various attention abilities that are addressed in the following remarks can also clearly be seen as touching upon all four perspectives and dimensions. In Chapter 1, outcomes of attention research in the Shamatha Project have been addressed. Attention, in Western approaches, has been described as the cognitive process of selectively concentrating on one aspect of the environment while ignoring other things. Attention is also referred to in terms of the allocation of processing resources. While clinical models of attention often do not coincide with academic investigation models, they both contribute much to our understanding. On the one hand there is, for instance, the academic investigator's "lines test," that we did for the project research, as described in Ch 1. On the other hand, a clinical practical hierarchic model for attention, described by cognitive rehabilitation experts seems relevant to me, also in the context of meditation. The clinicians mention five attentional abilities, in order of growing difficulty:

1. *Focused attention*: the ability to respond discretely to specific visual, auditory or tactile stimuli.

2. *Sustained attention*: the ability to maintain a consistent behavioral response during continuous and repetitive activity.

3. *Selective attention*: the ability to maintain a behavioral or cognitive set in the face of distracting or competing stimuli. It incorporates the notion of "freedom from distractibility."

4. *Alternating attention*: the ability of mental flexibility that allows individuals to shift their focus of attention and move between tasks having different cognitive requirements.

5. *Divided attention*: this is the highest level of attention, in the authors' views, and it refers to the ability to respond simultaneously to multiple tasks or multiple task demands.[1]

So, this is a cognitive rehabilitation sequence. Let's be honest, isn't that actually what we start from, in meditation: some sort of cognitive rehab? The cognitive components involved in attention, memory and executive functions overlap and interact in complex ways. The circuitry and structures involved are widely shared. The "executive system," based in the frontal cortex – the forehead-brain-cortex – controls our thoughts and actions to produce coherent behavior. It supports the maintaining of behavioral goals, and it supports using these goals as a basis for choosing which aspects of the environment to attend to and which action to select. As is emphasized, impairments in each of these five cognitive processes can have devastating effects for people in their daily functioning. Clearly, in this way of looking at attention in clinical context all four integral perspectives present themselves. Next to Behaviors, Experiences stands out, particularly in connection with what has been said about devastating effects on people's daily life.

It is my belief that all of the named five attention aspects are cultivated in our shamatha attention practice approaches. We train focused attention, sustained attention and selective attention with all the shamatha "tastes" that have been described in the context of the Shamatha Project. Alternating attention is done with all the named shamatha practices to a certain degree as: with mindfulness of the focus of attention, also brief moments of introspection are practiced. Alternating attention is more explicitly trained in Settling the mind with the initial double focus: the space of the mind and what arises in it. In the early phases there is an alternation that we can have some influence on, like with choosing foreground or background. Also there can be a shifting of perspective, with alternating event-perspective and mind-perspective from which we observe. For divided attention, the ability to respond simultaneously to multiple tasks, we may say that this also connects with the double focus, in more subtle practice, when we can hold them both. So, divided attention is trained as one observes the space of mind and what arises in it, with at the same time non-grasping to thoughts, while not banishing thoughts. In more advanced practice, division will be increasingly more subtle, with fleeting mind movements, and more continuity and completeness in staying. To say it in the terms that Alan Wallace used, during our exchange hours in the afternoon in the shamatha retreat: this coincides with increasing numbers of ascertained moments of awareness. With training the attention, increasingly more of the named 600 pulse-like moments of cognition per second will become ascertained, in the sense that we are aware of them.

10.2 Meditation and baselines

Let's now look at meditation through academic lenses, in a way that is also clarifying in exploring shamatha. For those studying meditation, three general issues are said to present themselves: 1. The claimed production of a distinctive and reproducible *state* that is phenomenally reportable, 2. The claimed relationship between that state and the development of specific *traits*, and 3. The claimed *progression* in the practice from the starter to the very experienced meditator.

All three issues, of course, have their body-mind and body-brain (neuro) correlates. In addressing these aspects I connect with descriptions by Antoine Lutz, John Dunne and Richard Davidson.[2]

1. Production of states that are phenomenologically reportable

Here, reporting can be about coarse, subtle, and very subtle states. It has been shown that the capacity of long-term practitioners to examine, modulate, and report their experience can provide valuable contributions to neuroscience research. High refinement can be developed regarding the range of experiences, as well as the abilities in detailed reporting.

As to shamatha, I wonder about the impressive changes in body-mind states that are described for a person who attains substrate consciousness and accomplishes shamatha. In the words of Alan Wallace (Chapter 4.3): this involves a radical transition in your body and mind. One is like a butterfly emerging from its cocoon, one experiences mental pliancy, and physical and mental bliss. The mind gets fit and supple like never before. There must be a shift in one's nervous system, and one may imagine that something remarkable must be taking place in the cortical regions of the brain; however, as Alan remarked during retreat, so far no one has monitored the brain correlates of this shift with an MRI scan or EEG. As far as I know, this still is the case.

More about states: it has been established that there is an on-going communication between the brain and the periphery, in both directions, and that this communication proceeds along three basic routes. First, there is *the autonomic nervous system:* this is the part of the nervous system that is involved for instance in regulating digestion, heart rate, respiratory rate, mostly unconsciously. Second, there is *the endocrine, hormonal system;* and third, *the immune system*. In each of these systems, specific pathways and signaling molecules make this communication possible. Looking back in my diaries, I can find traces of all these aspects that transpire in the diary presentations. I give phenomenal reports on contraction-relaxation states in the body-mind, cold-hot sensations, skin rashes. All the named three routes have been included in the assessments that were done in the Shamatha Project, while we were measured, monitored, and tapped into, for blood and saliva! Like, respectively: skin resistance for the autonomic nervous system, Oxytocin for a hormone, and Cytokine IL-6 for the immune system. All correlate to our meditative states. Reproducible and reportable states: this is a field where much will be won with increasingly more momentary state-measurement and reporting in the moment.

2. Relationships between states and specific traits

This aspect is part of the Shamatha Project's research questions, and of the hypotheses I posed about the shamatha practices. While state, globally, refers to a more momentary way of being, trait refers to more lasting dispositions and level of development of a person, as has been addressed in Chapter 8.5. The way Lutz and colleagues speak, in this context of meditation research, of *"transforming baselines,"* sounds appealing to me. Transforming the baseline state of experience means that the distinction between the meditative state and the post-meditative state is obliterated. When we practice shamatha, it's of course not intended that we are less distracted and less attentionally scattered, merely on the cushion: we aim to be more focused, also off the cushion. This may be seen as: *state into trait,* connecting state training with attentional and psychological stage progression We reach a new level, for instance on a line of the ability to better concentrate. Many studies have shown that at least several subcomponents of attention are best regarded as the product of trainable skills. Research regarding shamatha focused attention meditation has shown how these practices of attention and emotion regulation and non-reactive monitoring lead to a decrease in emotionally reactive behaviors: states into traits, with new baselines. As to the Shamatha Project: follow-up research has made clear, for instance, that we participants continued to show improvements in perception, and continued to show a greater wholesome capacity in response inhibition (connecting with less reactivity), as described in the articles by MacLean and Sahdra and their colleagues, referred to in Chapter 1.5.

One important aspect keeps intriguing me, also in this state-trait context: this regards introspection in meditation, meta-cognition, meta-awareness, the reflexive function. Alan Wallace describes meta-cognition as the knowledge of one's own cognitive and affective processes and states, as well as the ability to consciously and deliberately monitor and regulate those processes and states. He sees this phenomenon as a rich area for collaborative research, for contemplatives, scientists, and psychologists. Daniel Siegel, medical doctor and neuroscientist in the field of Mindful Awareness research, distinguishes between meta-cognition and meta-awareness. He describes reflection on the nature of one's own mental processes as a form of "meta-cognition," of thinking about thinking. Meta-awareness is awareness of awareness. I'm struck by one finding that John Teasdale and colleagues mention, in regard to their research on mindfulness based cognitive therapy for persons suffering recurrent depression. The psychological variable most associated with the increased resistance to depression, they found, is metacognitive awareness, the shift toward experiencing negative thoughts as observable mental contents rather than the self. Metacognitive awareness can therefore be called a crucial state-trait factor.[3]

Well, the shift from identification with thought to seeing thoughts as observable events – this is what conventional self-sense can handle, with some training. With contemplative practice, like Settling the mind, we go further: we may then proceed to non-grasping to thoughts, with melting of self-attachment, into more subtle states.

To phrase the state-trait question in a more general and expansive form: does waking up coincide with growing up? Does greater presence, by way of trained states of consciousness, coincide with greater human maturity, in the ability of seeing things from more perspectives? Yes, scientists confirm: with repeated practice of contacting higher states, your stages of development will tend to unfold much faster and more easily. One's state of consciousness may unfold in a way that no longer one interprets oneself merely as a gross ego, but experiences oneself as a subtle soul ...[4] This all correlates with less egocentricity, increasing sensitivity and increasing compassion. It appeals to me, not only in the context of meditation. With all kinds of mind, emotional and psycho-physiological training, with the inclusion of meditation: this unfolds a wide ranging evolutionary agenda. In Integral parlance: shifting from ego-centric to group-centric to world-centric view, to planet-centric view; with accordingly more mature and wholesome action. Increasingly opening our hearts!

3. *Progression from starter to experienced meditator*

When looking at the second point, state into trait, we can't avoid also looking at progression. In the context of shamatha, including Settling the mind practice, a trajectory has been described, with clear markers on the (Elephant) path that illustrates progression. We can describe this in an experiential way, and also, this progression has been measured, with attention research correlates. In relation to body-brain, body-mind, and progression, here I like to mention some special unique findings from the Shamatha Project. As referred to in Chapter 1: a connection as found of meditation and positive psychological change with changes on the DNA level. Those who did not meditate did not show the thirty percent increase in telomerase. Participants with the highest increase in telomerase showed the highest correlations with decreased neuroticism in the tests, and a greater sense of perceived control. These themselves were due to changes in self-reported mindfulness and sense of purpose. I think we may parallel this with "a progression from starter to experienced meditator"; coinciding with descriptions of a stunning progression in sense of well-being.

In integral terms, this section has mainly addressed meditation and baselines, in a general third person mode, complemented with first and second person remarks. The rest of this chapter will first bring attention to the intersubjective second person dimension. Included is a Compassion meditation. After that, more specific connections will be made between third person dimension with first person views, showing how fruitful the integration of various perspectives on meditation can work out for our understandings.

10.3 Crucial Heart Qualities

Contemplative training in the Four Qualities of the Heart presents examples of mind-emotion practices that have effected in more friendly and loving behavior, for practitioners through many centuries. These practices had an important place in the Shamatha Project. While until recently there was mainly the first and second person experience aspect regarding these qualities, now third person neuroscience research shows quite specific correlates. For instance, research has shown that maternal and romantic love in humans have been linked to the activation of specific reward and attachment circuitries.[5] For clarity: the term "attachment" refers here to ways of bonding and connection; the meaning of this term in this science context differs from Buddhist meaning (for some remarks, see Chapter 5, nr 19).

Research suggests that emotional and empathetic processes and prosocial responses are flexible skills that can be enhanced by training, and that with this training demonstrable neural changes are found. So indeed, the love we feel phenomenally shows in our neuronal network correlates. Like naming Positive Psychology (you might say: starting from the Experiences, first person perspective), this is the place to also name Positive Neuroscience, a movement that initiates research aiming at a better understanding in how the brain enables human flourishing. An "enriched" and inclusive third person Behaviors initiative!

As the Four Qualities of the Heart and shamatha attention practices do not demand allegiance to any religion or philosophy, they have been embedded in various secular programs, that teach attention and emotion regulation. We find an example of an approach with already many years experience in the Cultivating Emotional Balance project. This secularized program is designed to teach and evaluate the impact of meditation combined with emotion regulation strategies training on the emotional lives of beginning meditators. It focuses on teaching skills to better deal with destructive emotions. Recent research findings, in a setup with schoolteachers, show that this training activates cognitive networks associated with compassion, reduces negative emotional behavior and promotes prosocial responses.[6] Regarding the heart, intersubjectivity, and meditation research: Clifford Saron, Scientific Director in the Shamatha Project, makes some remarks in this context that touch me. One could say that he cautions against contextualizing the objectively measured effects of meditation too much in a direct cause-effect relation. Saron emphasizes how one aspect of the research is: " ... to actually consider how these are "baby steps" and how preliminary and immature this field is – and its maturing actually moves in the direction of trying to think about all the factors that might contribute to observed effects, and some of those may have nothing to do with meditation *per se*. They may have to do with factors such as social support and group dynamics, environmental effects in terms of your change in behavior, and your relation to a teacher."[7] The support of the group, the relation with the teacher: that sounds like very interesting possible additional specific research in a second person intersubjective context. These aspects may not be the direct stuff of focused attention training, however, they are crucial complementary factors in the meditational landscape, and in that way, in meditation research. Now it's time for a guided meditation, relating to interdependence and intersubjectivity.

Compassion, expanding the scope – guided meditation

This is a compassion meditation, inspired by some elements Alan addressed during retreat and also in his book *The Four Immeasurables*. In this guidance we cultivate compassion for a person who suffers, a dear friend, a person more "neutral," for yourself, for a person engaging in harmful actions, and up to all sentient beings. We are all interdependent. We train shifting toward more inclusive perspectives. Ego-centric to group-centric to world-centric to planet-centric ...

starts with yourself and
;in by bringing to mind a
adversity, physically or
the person to mind as
tuation this person is in.
arise for this person to be
fering ... Then apply this
neutral person ...
any suffering you want to
s, any sources of distress,
gs that you fear? Do you
is quite likely that your
ld like to be free of that.
to be free of suffering
re talking about and then
suffering. Like I wish for
myself, ... you be ... of suffering.

(*Regarding applying the practice toward a hostile person, there is the following invitation*). Bring vividly to mind a person who, as far as you can tell, does engage in actions that cause harm, whose mentality is afflicted with qualities such as ill-will, jealousy, spite, or selfishness ... Now briefly bring your awareness back to yourself and imagine what it would be like if you yourself were afflicted with such a disposition, with similar habits of behavior. It may feel like your horizons are shutting down, your world is growing smaller, your heart becomes contorted ... Yearn to be free of these afflictions of the mind, unencumbered by such behavioral tendencies. Restore yourself to more spaciousness

> and light, and imagine being utterly free of those tendencies ... Turn your awareness back to this same person, and let the yearning arise, "Just as I wish to be free of such afflictions and harmful behavior, may you also be free." See the person who is afflicted, without equating this person with the temporary afflictions of personality and behavioral patterns ... "May you indeed be free of suffering. May you find the happiness and well-being that you seek. May all the sources of unhappiness, pain and conflict fall away. May you be free of suffering and its sources." Imagine this person as vividly as you can, free of those sources of suffering.
>
> (*Then Wallace invites you to extend*). Now expand the scope of this compassion to all sentient beings in each of the four quarters, attending first to the reality that each one essentially wishes to be free of suffering. It is this yearning that accounts for such very diverse behavior, some of it wholesome, some of it terribly harmful. Let your heart be joined with their essential yearning. "May you indeed be free of suffering, just as I myself wish to be free of suffering." Let your body fill with light and send it out to each of the four quarters. Imagine all humans, and then all sentient beings in each of these regions emerging from suffering and the sources of suffering.[8]

The theme of interdependence and the notion of intersubjectivity are central to all of Buddhism. In the Buddhist view, as we saw, each person does exist as an individual, but the self or personal identity is not existing as an independent ego-personality, that is somehow in control of the body and mind. Wallace emphasized that our existence is invariably intersubjective, as we exist in causal connections. We are constantly influenced by, and exert influence upon, the world around us, other people included. Given this, training compassion for self is, in the end, no different in its effects than training for others.

10.4 Research on two basic types of meditation

It has been emphasized that Settling the mind meditation is a shamatha calm practice, not a vipashyana insight practice – while it became also clear that the practice brings in surprising insights in the psyche. Interestingly, Western researchers have operationalized two basic meditation styles, based in Buddhism, for doing scientific research: Focused Attention meditation, and Open Monitoring. In my view, they can clarify some aspects of Settling the mind practice in a larger research context. Also they may offer a basis for operationalizing future research on the Settling the mind practice. In this, a number of publications by Antoine Lutz, Richard Davidson and colleagues from recent years, and some others, form the thread. Meditation, in this research terminology context, is conceptualized as a family of complex emotional and attentional regulatory strategies developed for various ends, including the cultivation of well-being and emotional balance. [9] Following are descriptions of Focused Attention, and Open Monitoring. With some variations, both styles are found in forms of meditations that are most familiar to Westerners: Vipassana, Zen, and Tibetan Buddhism.

Focused Attention

The approach involves sustaining selective attention moment by moment on a chosen object, in that way cultivating the acuity and stability of the attention. Connected with this, the practice develops three skills, that we came across, phrased here in objective research language. The first is: the monitoring faculty that remains vigilant to distractions without destabilizing the intended focus. The second is: the ability to disengage from a distracting object without further involvement; and the third, the ability to quickly redirect focus to the chosen object.

In the Shamatha Project, attention practices have been done with an intentional structure of subject and object, belonging to the research category of Focused Attention. The Focused Attention techniques include both the one-pointed attention shamatha, with techniques that cultivate a form of voluntary, effortful and sustained attention on an object, and vipashyana meditation. Vipashyana aims to cultivate a more broadly focused, non-judgmental mode of "bare attention."

Open Monitoring

The researchers refer to "Open Monitoring" as a way of functioning, and to "Open Presence" as a state of consciousness, a state of being. Open Monitoring practices share a number of features, including in particular the initial use of Focused Attention training to calm the mind and reduce distractions. Some stability first! As Focused Attention gets better, the well developed monitoring skill becomes the main point of transition into Open Monitoring practice. The meditator then aims to remain only in the monitoring state, while being attentive moment by moment to anything that occurs in experience. In the instructions most of the time increasing emphasis is given to cultivating a *reflexive awareness* that offers greater access to the rich features of each experience, like the active cognitive dynamics, the measure of phenomenal intensity, and the emotional tone. Thus, as Lutz and colleagues say, unlike in Focused Attention, in Open Monitoring there is no explicit focus and there is no strong distinction between selection and de-selection. This "reflexive awareness" has also been addressed in Chapter 7.2.

Connecting to the Shamatha Project meditations

Now, let me connect these notions with the Shamatha Project practices for a little exploration. In the project breath practices we make a clear selection. While I may at first select focusing on the tactile sensations of the breath in the abdomen as the object, later I select the focus at the nostrils; in that case the focus on the abdomen is "de-selected." We practiced shamatha meditations, mostly in the sense of attention practices. It is also noteworthy that the Four Qualities of the Heart are named shamatha practices, as they have a focus; they can be named "discursive shamatha practices," as they directly involve the thinking mind, including words, imagination, and emotion.

Settling the mind, while including the space of the mind as an object, is a practice of selectivity. While Open Monitoring gets more and more open and all-encompassing to all that offers itself, Settling the mind shamatha keeps selectively limiting itself to the mental. So, with Open Monitoring, one may say, the "effortful" selection or subtle grasping to an object as a focus, as done in Settling the mind, is replaced by the "effortless" sustaining of awareness, openness, without selection. Emphasis is on reflexivity. There is great stability in Open Presence, in the sense that one is not easily perturbed out of this state. Interestingly: in shamatha, stability is constituted by the fact that other phenomena do not pull you away from the selected object of mindfulness. Instead, with Open Presence, your stability consists in the ability to continue to experience phenomena without objectifying them and, ideally, without having a sense of a "doing" subjectivity.

Shamatha and some neural correlates

Following are some terms you may come across, reading in the field of attention and meditation research. While some locations are named, this is not so much about structures, as well as dynamic, ever-changing neuronal association networks. Mention has been made of the executive system and frontal cortex. To name one brain-location correlate, crucial in this context of concentration meditation: it has been found that for allocation of attention during competing attentional demands especially the *anterior cingulate cortex* of the brain (part of the prefrontal cortex, located behind the forehead) is important. Rick Hanson and Richard Mendius, neuropsychologist-meditation teacher, and neurologist, in their book *Buddha's Brain*, present some interesting connections. The prefrontal cortex is involved in setting goals, making plans, directing action and shaping emotions. While reading in this context, you may also meet the insula, involved in sensing the internal state of your body (introspection!), including gut feelings, and also helping you to be empathetic (The Four Qualities of the Heart). Some other important structures are: the hippocampus, connected with forming new memories, and detecting threats; and the amygdala, described as a kind of alarm bell that responds to emotionally charged or negative stimuli. The limbic system, central to emotion and motivation, includes the hippocampus and amygdala – so this certainly is crucial also in relation to The Four Qualities of the Heart. And let me also name here the brain stem that is involved in sending neuromodulators such as serotonin and dopamine to the rest of the brain. High transmissions of dopamine are connected to joy and bliss in concentration meditation; they are said to keep the gate to working memory shut so you can become increasingly absorbed with inner life. Another aspect, relevant in shamatha context, is that single-pointed attention is probably supported by fast gamma wave synchronization of large areas of the brain.[10] Oh, I remember how my imaginations, in diary fragments, went that gamma way (Chapter 5.3, nr 25). This is just very coarse correlation naming, while the crux is in the moment-by-moment dynamics, in sprouting and pruning, in rewiring, in connections, networks, facilitating, inhibiting, and dis-inhibiting dynamics, and much more.

More trust: acquired secure attachment style

There's one aspect in this sense of dynamics and their correlates that really speaks to me. In the diaries, fragment (19) in Chapter 5, in describing experiences in ongoing meditation, I name how I feel "more trust, more safe internally," more at home, familiarizing. With that, I refer to attachment style, and a sense of shifting to a more acquired secure attachment style. "Attachment style," primarily developed early in life, reflects the way that a child is responded to and how it interacts with primary caregivers. There can be more or less attunement by the caregiver, which generally results in later styles of being and relating, resonating in a more or less secure way of relating with others. The less secure may involve a dismissing style, a preoccupied style and it may include dissociative tendencies. Research has shown that, even when attachment style – in psychological sense – is not eminently secure, there can be a change during later life. This then is called acquired, or earned secure attachment. I feel excited, reading about research in this area that connects with meditation, as do some remarks by Daniel Siegel. In connecting the fields of attachment and attunement with mindfulness practice, he proposes that the experience needed in childhood via attachment with parents can be paralleled to some degree with the internal attunement of mindful practice. Mindful practice is said to reinforce adaptive self-attunement and regulation. I recognize what he says: we develop a secure attachment with ourselves, in mindfulness practice. Siegel posits that we can then make the link that this form of internal attunement would also promote the healthy activation and subsequent growth in the same social and self-regulatory prefrontal regions that have been involved in early development. Referring to this as the "mindfulness-attunement hypothesis," he suggests that attuning the mind to its own mental processes is the essential feature of mindful awareness practices. Seven functions of the middle prefrontal region are named as being associated with both forms of attunement: body regulation, attuned communication, emotional balance, response flexibility, empathy, self-knowing awareness, and fear modulation. In this sense, we may indeed say that with certain meditation practices we are developing a secure attachment with ourselves – we come to trust that we can rely on ourselves for comfort and connection. So, this is another example of shifting baselines, from states to traits, in a more wholesome and encompassing way. The less dust, the more trust!

Interestingly, when immersing myself in articles about research in these areas, I often meet with tests that sound familiar, tests that we did in the Shamatha Project, like kinds of the Visual Continuous Performance Task, as was the "lines test" that we did. Also the Stroop Test and the Posner Cueing Test pass by (see 1.4).

Diary after retreat, 2011

> Reading about meditation in these third person terms reminds me of the different felt sense of doing the "lines-test" myself: to start, on the one hand, first person sitting with the computer, in this dark little room. The detailed refined measurements, eyes vigilant on the lines, fingers intent on the button ... years ago. And then more recently, on the other hand, reading the article by Katherine MacLean and her group about this test in a professional journal, third person language. Which in some sense felt also reassuring, in giving a larger context to our little lines, in the larger project, and in the broader perspective of ongoing research projects. Shifting perspectives ...

As was referred to in Chapter 1.5: it's a wonderful development that quite some neuroscientists increasingly take first person report very seriously. Different meditation practices will connect with different neurophenomenological descriptions and with different neurophysiologial measurement outcomes. More assessments of psychological changes, in their state-trait aspects, and of state-trait neuroactivity markers across various meditative practices, will be necessary for seeing their potentials. This will include developing and investigating their clinical utility. Verbal reports about inner experiences can easily be biased. They are personal. It is crucial that precise and rigorous first person methods are developed, that, I think, account for and honor personalness. Long-term practitioners can generate more stable and reproducible mental states, and can describe these states more accurately, with more refinement, than beginning meditators. Fortunately, many Buddhist contemplatives have reported their findings through the ages, in very detailed ways as described by authors through the centuries. Given some understanding of their perspectives and methods, their findings in contemplative practice offer rich insights in what body-speech-mind may bring forth, as we saw with the descriptions by Düdjom Lingpa. Also more recent examples of detailed descriptions of meditation experiences can be very clarifying.[11]

For proceeding on this path we will need standardized methods and protocols. The following protocol to me seems very helpful for systematization in research, and for communication between first, second and third person. How might this work for Settling the mind in its natural state practice? Below, I make an attempt to fill it in. Hopefully, future research on Settling the mind practice will use a protocol like this one, completed by a group of practitioners.

10.5 A model protocol for meditation research

Antoine Lutz and colleagues have provided a model protocol that can support preciseness in the dialogue between meditators and researchers. This can lead to better chances for fruitful interaction between those in neuroscience and in meditation.[12] Making things more explicit is certainly not a luxury, as in communication about various meditation tastes there can be a lot of confusion about who means what. Through centuries and through cultures, significant changes in meditation styles are found, even for a basic shamatha style of meditation focused on the breath. Here is an example regarding the eyes: in the vipassana movement one may be advised to close the eyes, so as to de-emphasize the importance of the visual sense-modality; while closing the eyes is rarely encouraged in Zen and Tibetan traditions, in part because it might induce dullness or drowsiness. All parties have their valid arguments. Another example lies in assessing the degree of training: quantification of a total of hours of meditation throughout a practitioner's life doesn't work in a straightforward way, as many varieties of meditation mutually influence each other and differentially affect the mind. So, let's say that you have performed 100 hours of vipashyana meditation, and your friend has practiced 100 hours of vipashyana preceded by 50 hours of shamatha. These two vipashyana stretches would not be the same, and this needs to be accounted for in the outcomes.

In this protocol, an attempt is made to cover qualitative

meditation practice aspects. It describes a first person experience perspective, to be correlated with a third person perspective. When I read it, I felt inspired, and invited. Five basic categories of questions are provided, that address 1. The relative degree of stability and clarity appropriate to the practice; 2. The intentional modality, referring to whether the meditation has an object; 3. The techniques, such as breath manipulation, that are employed; 4. The expected effects of the practice during meditation; and 5. The expected effects after a session.

Settling the mind in its natural state, a structured description

In this section you find the protocol outline, with my tentative presentation of the named aspects in the way that I experience these for Settling the mind practice. I invite you, having some experience with Settling the mind meditation now, to first read the question and cover my response with your hand. See what comes up for you.

1. Concerning stability and clarity:

1-1 In view of the practitioner's level, should the meditation favor stability, clarity, or a balance?

Settling the mind in its natural state is a form of shamatha meditation, often described as the shamatha practice in Mahamudra and Dzogchen. This meditation addresses quite a range, as to the named qualities: first stability (that rests in relaxation), then adding clarity; all the time there is introspection and monitoring for balancing the three. In the course of progress, the balance is shifting.

1-1a What are the indications that stability needs adjustment?

Excitation and laxity indicate that stability needs adjustment. In this mind context, also distraction and grasping show that stability needs adjustment.

1-1b What are the indications that clarity needs adjustment?

Especially, dullness and laxity are indications.

2. Concerning intentional modality:

2-1 If the meditation includes an object, then:

2-1a Is there one object or many objects in the meditation?

This meditation is mental object focused. Object of meditation is, initially, a dual object: 1 - the space of the mind, and 2 - what arises in it (and abides and dissolves in the space of mind).

2-1b For each object, is the object dynamic or static?

Regarding 1 - The space of the mind may be experienced as "dynamic," in its aspect of individual mental space. This may shift into a larger space, the space of awareness. The space may be experienced as lively, and at the same time, possibly: changeless.

Regarding 2 - What arises in it is experienced as dynamic.

2-1c If the object includes or consists of a visual form, a sound, or a sensation, then is the object perceived through the senses, or is it imagined in the mind through visualization or another technique?

The objects are mind-objects; the dynamic objects include thoughts, emotions, plans, memories, imaginations (for instance, visualizations). So, the objects are experienced, perceived, and imagined in the mind in a number of ways.

2-2 If the meditation does not include an object, then does one direct one's attention to something else?

The meditation is object-focused.

3. Concerning meditative techniques:

3-1 Is the practice done with the eyes opened or closed?

The practice is done with eyes open; with vacantly resting one's gaze in the space in front.

3-2 Does the practice employ any discursive strategies, such as recitations, memorized descriptions or arguments that one reviews?

The practice employs the basic discursive strategy, in the beginning, of learning the instructions, then thinking about the practice. Then there is a shifting to decreasing discursiveness, mindfulness, introspection ... (see Wallace, 2006: 174-5). The instructions include knowing for oneself what are the objects of meditation (2-1a), and then, to observe these objects, in non-distraction and non-grasping. This means, initially, observing space of mind and thought, while releasing grasping to thought. Later there is familiarity with the on-going practice. Then discursive strategies are no longer needed.

3-3 Does the practice use breath manipulation?

The practice does not use breath manipulation, in the original description as given by, for instance, Lerab Lingpa and Düdjom Lingpa. In the instructions by Alan Wallace, in the initial phase, the possibility is given to connect this practice with (spontaneous, not manipulated) in and out-breath in a specific way, of releasing grasping, paired with the out-breath.

3-4 Does the meditation involve focusing on different parts of the body, by means of a visualization or some other technique?

No.

3-5 Does the practice require a specific posture or set of physical exercises?

The practice requires the accepted basic meditation posture(s), with the detail: eyes (and possibly mouth, a bit) are open.

4. Concerning expected effects during meditation:

4-1 Is the meditation expected to produce any physical sensations or mental events, either constantly or intermittently?

Physical sensations are not intentionally "produced" by the practice, but they will come up and the practitioner lets go of these, as they are not the selected object(s). The practice does not produce sensations or events: there is mindfulness, selective observation of space of the mind and mental events, of what's present. Mental events are observed (in their arising, being and dissolving), and not grasped at. With practice, mindful attention will be increasingly continuous on the object, and complete; in the sense of less partial. Events will gradually be more subtle fleeting mind moments, up to very subtle "union of stillness and motion."

In initial phases, there is intermittent introspection and monitoring, for quality control. In advanced phases there will increasingly be a more continual meta-awareness.

So, intermittent are the arising, being and dissolving of mental events, in the space of mind. While the arising of (increasingly subtle) mental events is expected to decrease to practically none, meta-awareness (also referred to as reflexive awareness, and awareness-itself) will be more and more prominent and constant, with a muted sense of duality of subject and object.

4-2 Does one expect the meditation to produce subjectively noticeable alterations in cognition, either constantly or intermittently? One example would be the impression that one's perceptions seem to be like the appearances in a dream.

Yes, the meditation is expected to "produce" – in the sense of: lead to – alterations in cognition, with a great variety. In the beginning phases, cognitions will be connected to coarse states of awareness. There may be transient anomalous physical and mental experiences, catalyzed by the practice. With increasingly stable and vivid attention, and more subtle states of awareness, cognition changes, and "cognizance" is experienced. Cognizance refers to the knowing quality of awareness. With decreasing grasping there will be changing self-cognition, with decreasing self-attachment.

4-3 Is the meditation expected to cause any emotions, either constantly or intermittently?

There may be phases in the meditation process in which, possibly, strong emotions may come up, transiently, intermittently, in great variety of quality (angry, anxious, joyful, blissful) and quantity. Progressing further with the practice, more and more, emotions are "tamed." Then there will be calm, with emotions and involuntary thoughts "like a river slowly flowing through a valley," to calm "like an ocean with no waves." This may go up to the conceptually discursive mind being still like

Mount Meru, King of Mountains – in the words of teacher Alan Wallace. With increasing refinement in training, more often a sense of joy, luminosity, nonconceptuality and "cognizance" have been described by practitioners.

5. Concerning expected effects after meditation:

5-1 Does one expect the meditation to alter one's cognitions? One example would be the impression that one's perceptions are more vivid.

Yes (see also 4.2), cognitions are expected to change over time: increasingly more vivid, clear, with a mind more calm, supple, and limpid. Increasingly there will be less identification with thought and with a sense of separate-self. In cognizing there is less effort, there is increasing effortlessness. In this specific shamatha practice of Settling the mind in its natural state, many insights in the psyche come up, like around contents, patterns in psychological dynamics, and also regarding the nature of the psyche. Cognizance is experienced.

5-2 Does one expect the meditation to alter one's behavior? One example would be a tendency to sleep less.

Yes, as in many forms of shamatha meditation, there may be a tendency toward less need for sleep. The effects as described in 5-1 for cognitions will also have their effects in behavior: often there will be more mindful awareness, less reactivity, more presence in whatever is felt, thought, done, lived, with others and oneself, and the world. Often, meditators change their lifestyle toward a more simple, sober, contemplative life, with more psychological, spiritual, existential interests, and more compassionate action toward others, the world and themselves.

5-3 Does one expect the meditation to alter one's emotions? One example would be the tendency to recover more quickly from emotional disturbances.

Yes, the expectation is that emotional life will change; with less grasping to emotions, less identification with emotions, less

> emotional turbulence and disturbance, and quicker recovery from these. There will be more positive emotions, like empathetic joy, and the emotional ground tone will overall be more one of ease, joy and equanimity. While they cannot be called "emotions," after prolonged practice, experiences of bliss and luminosity are described.

Well, if you have first seen for yourself what responses came up to these questions, did they match with what I wrote down?

10.6 Two shamatha pilot-explorations, some impressions

In the explorations concerning Settling the mind in its natural state the notions mind, settling, natural state are mentioned quite often, as can be expected. This is done with an apparent assumption that we all have an idea what is referred to with these notions. Probably, we assume that we are in some agreement as to what these notions stand for. In what follows I present some empirical impressions that include giving nuance to this assumption.

Out of curiosity, I carried out two questionnaire-based shamatha pilot-explorations, following the Kalama Sutta's invitation: look for yourself! Trust your own heartfelt interest and follow it! May I add: even if it's done with very basic tools.

The first, the "Settling the mind questionnaire," directly connects with Settling the mind practice: it is about the ideas that starting Settling the mind practitioners have about *mind, settling,* and *natural state*. For this first person understanding approach mostly multiple choice questions have been used.

The second, the "After shamatha retreat questionnaire," relates to the way participants in a two-month shamatha retreat look back on that experience, and how others experience these persons. This "others" aspect, often missing in research, in my view is of crucial interest: how is the person, having participated in a retreat, perceived by others? Has she changed, toward being more cooperative? Is he more trusting, more seeing the needs of others? That's where the real fruits must lie, in one's conduct in the world! Here Likert scales have been used for obtaining subjective ratings. Both of these empirical explorations included third person quantitative measurements. While these investigations obviously have their methodological flaws, the responses to the questions present valuable information. The responses may not lead to brand-new understandings of the practices, yet, to my feeling they add illustrations, sometimes surprising and touching, from the angle of how the practices and the mind are understood; and how intensive practice for a period of time works out, for persons "like you and me."[13] Here are a few impressions.

"Settling the mind questionnaire"

While guiding persons in shamatha meditation courses, I've heard quite a few comments and questions about mind. This of course can be expected, given the various uses of the terms, including as addressed in Chapter 7.2. We can just choose a definition, which helps to be on one line while sharing. However, I thought, there may be an interesting step preceding that, which is to investigate with what kind of ideation about mind participants start a shamatha course. So, at the beginning of the first meeting, without much talking about mind, the participants have completed a brief questionnaire, and this was repeated at the end of a six-week course.

As mentioned earlier, in guiding us during the Shamatha Project in Settling the mind in its natural state, Alan Wallace included some bits of "mind-vipashyana." We did this in addition to the familiar moments of introspection-monitoring. This was a brief period of practice of, what I would call, sensitizing ourselves to the mind. Wallace invited us to questions like: what is the mind, and where is it located? Does it have a shape, a color, a texture? Does it have a center? A few of these have been included in the questionnaire.

The questionnaire was completed by fourteen mental health practitioners, participants in a shamatha practice course. They had three course days, and practiced one hour meditation a day at home, during six weeks. Pre and post-measurements were made. The application of the questionnaire had a twofold aim: 1. Offering to the participants some structured practice in sensitizing themselves to mind, and 2. Giving an impression of how, in the pre and post-scoring, the participants report their experiences and understandings relating to mind and the Settling the mind practice.

Where is the mind, who does the settling?

Regarding the question "Where is the mind?" more than 70 percent of the participants reported thinking that mind is both in, outside, and beyond his or her head, at both pre and post-scoring.

For questions in the sense of: "Does the mind have a center, a periphery, boundaries? Does it have a color, a consistency?" we see a clear shift in reported understanding, with the trend to less of all: less center, periphery, boundaries, color and consistency, in post-scoring, as compared to pre-scoring.

For the question "Settling the mind in its natural state: who or what does the settling?" response categories are: I do it, The mind does it, Settling just happens, and Other. There seems to be an increasing awareness of one's own initiation of the process, while consequently "Settling just happens"; this shows in the remarks that respondents added. For qualitative input: a few participants add some nuances to their scoring. One person, at the post-score, notes: "Settling just happens," complementing this with "but 'I' follow the instructions." Another person, at pre-scoring chooses "Other," adding "the mind and/or the soul," and at post-scoring chooses for the inclusiveness of "I do it, The mind does it, and Settling just happens."

Comments: interestingly, these qualitative remarks for me bring up the association that participants get to perceive in more refined ways in the practice-process. In some sense, this feels like them taking responsibility for their motivation and initiating the mind practice, that consequently "just happens" and evolves.

What might be "the natural state?"

For the exploratory many-item question "When you consider the 'Natural state of the mind' ... what comes up for you?" possible categories are: 1 What I experience or am aware of during my normal day (for instance, thinking, emotions, memories), 2 What I am aware of and what I am not aware of (for instance, unconscious), 3 Mind without content (for instance, the ground of the ordinary mind or psyche), 4 True Nature mind, 5 Openness, 6 Love, 7 Equanimity, 8 Other.

Quantitative outcomes show that the largest shift, in comparing pre to post-scoring, is in a decrease of 3 Mind without content, and an increase in 4 True Nature mind. Also 5 Openness finds an increase.

Comments: how striking, this shift to the reported understanding for the "Natural state of the mind" as True Nature mind! What is natural? We saw that natural state is used in different ways in texts, within different contexts. In some texts, indeed, the natural state directly relates to True Nature mind: " ... if you have some recognition of your mind's nature, then, when any one of those thoughts arises, you will experience the mind's true nature in that thought, because the mind's nature is also the nature of that thought," says Thrangu Rinpoche, in Mahamudra context (Chapter 7.5). At the same time, Alan Wallace's guidance has referred to the natural state, in terms of shamatha, as substrate consciousness: the relative natural state. To what degree do these findings relate to a person's experience, to what degree to guidance? The main thing has been: the invitation for a form of reflexivity, of being aware.

Summarizing a few findings:

This pilot-exploration has shown some shifts in reported experiences, in participants in a six-week shamatha course. *Mind* is understood in a broad way, increasingly without ascribed qualities (like having a center, a periphery, boundaries). Regarding Settling the mind in its natural state practice, there seems to be an increasing awareness of one's own initiation of the process, while subsequently "Settling just happens." As to what is conceived of as the "natural state of the mind," while at first "mind without content" is often reported, later the natural state is increasingly connected with True Nature mind. There is a shift to more subtle experiencing.

"After shamatha retreat questionnaire"

This second pilot-investigation has been made among participants in a two-month shamatha retreat, taught by Alan Wallace in 2010, in Phuket, Thailand. It includes a questionnaire that is filled in by the former participant, and by one or two near and possibly dear persons, at points in time around one month (T1), and five months (T2) after concluding the shamatha retreat. So, this exploration involves reports about how the participant experiences him or herself (first person subjective), and how the person is experienced by others (second person intersubjective), on a number of dimensions. "Others" (to be named this way) have included partners, friends, colleagues, and also a parent, and a teenage child. While this questionnaire is not specifically about Settling the mind meditation, outcomes include the effects of this practice.

The questionnaire starts with a few open questions, relating to the way that the former Participant experiences her or himself – and Others experience the participant – after retreat, in comparison with the situation before retreat. Reference is made, for instance, to contact, emotions, moods, attention, presence. Then, 16 questions are presented in the format of responses on Likert scales, with scoring -3 to +3 (ranging from "very much less" to "very much more," in comparison with pre-retreat). These 16 questions can be grouped in five clusters, being: (1) Attention skills (ability to focus, ability to stay concentrated for a longer time, to not forget what is the focus, ability for not being easily distracted, and "presence," awareness in the moment); (2) Emotion regulation (questions relating to: ability for self-reflection, expressing positive emotions, expressing negative emotions, ability to regulate emotions, quality of contact, capacity to handle conflicts with others). A third cluster regards: (3) Qualities of the heart, social interpersonal aspects, like: openheartedness toward others, love, compassion, taking interest in others; and the ability to reach out toward the other and being helpful. A fourth category was named: (4) "At homeness" with oneself: including being balanced, at ease, peaceful (for Others: the impression of this being so), and an ability to be clear to others, including setting limits if appropriate. Lastly, cluster (5) refers to "Being easily content," in the sense of content with relatively few possessions, and few external pleasures.

Eighteen Participants, with connected one or two Others, have been involved in the scorings. In total, 31 Participant scorings, and 28 Other scorings have been used in the analyses. Following are some findings.

How much, and how, do you and others see you as changed, after retreat?

When looking at the five clusters of phenomena (Attention, Emotion regulation, Heart-social, At homeness, and Easily content), outcomes make clear that, *on all five aspects, both Participants and Others see considerable improvements.* This is the case at both T1 and T2. At both assessments, for all clusters improvements are noted, in comparison to these abilities and skills before retreat. The mean increase in scoring, overall, when taking all assessments at T1 and T2 together, is +1.47, for both Participants and Others (scale ranging from -3 to +3).

While the improvements at T2 globally are somewhat less, compared with T1, they still are impressive, as compared to before retreat. When we take all five clusters together for the Participants, we see a decrease of mean value (from 1,62 to 1,32), which is statistically significant (p-value < 0.05). The decrease as seen for Others scoring (while not that much different: 1,61 to 1,33) is not significant. Clearly, the wholesome effects of the retreat are still experienced as quite strong, also 5 months after the retreat was over. Following are some more observations, illustrated with quotes by Participants and Others.

(1) Attention

After retreat, especially Attention is reportedly dramatically more focused and stable, particularly as noted by the Participants themselves (1.88). While 5 months after retreat this effect has somewhat decreased, attention skills are clearly experienced as much better than before retreat, by both Participant and Others

A Participant notes, at T1: "I still get lost in thoughts and emotions, but I have a stronger introspective faculty and am able to catch the wandering mind more often."

For Attention, at T2, a Participant describes: "More present, more resilience, more focus in life, more quiet determination in what needs to be done." One Participant presents an interesting comparison: "I am much more calm and patient. That has stayed about the same since my return ... My attention skills have dissipated somewhat since I returned, when they were very sharp. My calmness level and patience level have remained the same since I returned." Others note the following: "She has more capacity of attention, she has a better mood." And "He is more conscious about everything in his life and everyone around him."

(2) Emotion regulation

One Participant observes, at T1: "I am less maligned by negative thoughts. Realizing they aren't me, I am better able to let them go when they come up." Another notes: "I still get angry, but quicker than before I observe it, recognize it, and try to apply antidotes or focus on my breathing. At other times I can recognize it before I get angry and dissolve it and be free of anger!" One Other mentions: "One aspect where I can really see a difference is in trying to control her emotions and how to handle conflicts with herself ... she deals with it better than before." Also, an Other notes: "(the person) is more sensitive, better mood, less fear and anger."

At T2, one Other mentions: "Although diminished from her return from Phuket, (the person) still retains significantly greater equanimity, balance, and objectivity than before her retreat. She continues to be able to see and let go of thoughts and emotional reactions earlier before they can claim her."

Interestingly, for Emotion regulation, Others have an impression of more positive change than the Participants themselves (for both T1 and T2 scores combined)! For Emotion regulation, Others show a trend in estimating the abilities as better than the Participants themselves at both T1 and T2! Might the phenomena in this cluster be most visible, and relevant to Others?

(3) Heart-social

A Participant notes, at T1: "I experience myself as more contactful, and I became a better listener, I think."

Here is a trio combination, at T2: a Participant writes down "I feel more easy-going and kind to people around me"; a family member notes about this person "More at ease ... More compassionate and more interested in others," and a good friend mentions: "I find (the person) ... to be more patient and joyful. He ... is a lot more easy going and fun to be around."

(4) At homeness

The most stable, and largest positive effects are reported by the Participants in relation to (4), At homeness with oneself (at T1 + T2, the combination of these two being highest of the five cluster scores). Some quotes:

One Participant observes, at T1: "I think being in retreat has been the best time ever spent in my life and the most deep act of love with myself." Another Participant declares: "I feel more open and free: more spontaneous, playful, creative, confident, connected to others, focused, happy. This comes from an inner sense of well-being and self-acceptance that I didn't feel before the retreat." And another: "Altogether: more simple, sober, happy, open."

At T2, a Participant describes: "I feel very much in touch with myself in a constant way." Two Others state: " ... she is clearer about her interests and boundaries ... she has been calmer, more patient with herself," and "He used to be somewhat of a perfectionist, and now he is less hard on himself and able to relax and enjoy simple parts of life."

(5) Easily content

A remarkable finding for this aspect is that only in this one cluster, Participants describe a process of increasing, after retreat: the score is higher at T2, as compared to T1.

Some quotes are the following: at T1 an Other states: "She now wants to be in a more calm and quiet environment and is more interested in meditating or reading than in watching TV or just looking for some type of entertainment."

At T2, a Participant notes: "I am paring down my possessions to those few that have the most meaning for me ... Objects are no longer a source of satisfaction to me."

How much do you think these aspects have to do with participation in the shamatha retreat?

Regarding the role of the shamatha retreat: both Participants and Others emphatically connect the described wholesome changes in, for instance, attention skills, self-acceptance, emotional and social skills, to participation in the retreat. Regarding the crucial *Integration* in the daily life, back home, the place where abilities, skills, and possible changes are "lived," some more quotes can be presented.

At T1, one Participant states: "I feel very centered and able to deal with situations in a calm and relaxed manner."

A Participant at five months after retreat, at T2, finds: "I am more confident in myself ... more purposeful, more directed, more present ... I have little interest in acquiring new things except as needed ... Overall, I don't feel meditation has progressed much (in terms of stability-levels) since Phuket. Even so, the way I live in the world is profoundly changing. It is clear to me this has to do with meditation."

Summarizing a few findings:

Regarding five clusters of phenomena (Attention, Emotion regulation, Heart-social, At homeness with oneself, and Easily content), outcome-reports make clear that *both Participants and Others see considerable improvements, in comparison to before retreat.* This is the case at T1 (one month after retreat), and also at T2 (five months after retreat). Specifically, reported Attention is dramatically more focused and stable after retreat. While 5 months after retreat this effect has somewhat decreased, attention skills are clearly experienced as much better than before retreat, by both Participant and Others. The most stable and largest positive effects are reported by the Participants in relation to At homeness with oneself. Participants and Others emphatically connect the described wholesome changes to participation in the shamatha retreat.

A last remark regarding this exploration: I am impressed with the nuance with which both Participants and Others view the Participant in his or her more or less or not changed ways. Thank you so much, dear respondents.

And now ... Let's drop all effort, all discursive understandings, all theory, wonderings and musings, about practice, and context. Drop it all. Just sit and be. Relax, be open, receptive, with all the six senses. Be aware with what is. Who am I, what is mind, what is the natural state: this is it, right here, right now.

Chapter 11

Applications, integrations: shamatha in society

One aspect of executive function is the ability to make mental shifts and engage in flexible thinking. That rings a bell. The inability to do that was exactly what has been quite a confrontation ... (Diary after retreat, 2010, 11.5)

Let's now see what shamatha practice, and Settling the mind in particular, may offer in their applications, aimed at decreasing suffering and increasing resilience, in various kinds of situations in people, in the world.

It always starts with attention. Attention, the term derived from Latin (ad-tendere) refers to: tending, to tend, to take care of. Alan Wallace named his book that came out in 2006: *The Attention revolution*. It addresses the neglected place that attention training has had in our society. We need a revolution in the sense of climbing the barricades, claiming attention for attention! The revolution is about helping ourselves and others cultivate focused attention, for focused action, and certainly also it is about practicing the art of attention for others.

In this chapter I return to the hypotheses regarding Settling the mind practice, posed in the Chapter 3.6. After the exploration expedition in the preceding chapters, with a description of the project, the contemplative and scientific aspects, diary experiences, inquiring into Buddhist texts and contexts, with musings on some psychological and science issues: what has come out, what may be concluded, so far? Of course, there is some playfulness in this set-up. Science with a wink; still, serious playfulness. My first person n=1 study cannot bring in hard evidence, but at least, my experiences may add some extra personal illustration to robust research findings. With my comments to the hypotheses, I will refer globally to the main chapters where the topic has been elaborated. Next to that, possible applications of shamatha in general and Settling the mind practice in particular in society are addressed.

11.1 Back to the working hypotheses: observing our minds

Hypothesis 1, as phrased in Chapter 3.6, is: "Shamatha practices, and specifically the meditation practice Settling the mind in its natural state, provide first person phenomenological understanding of dynamics in the psyche."

Questions relating to dynamics in the psyche (the relative individual mind) have been addressed in presenting and exploring views from Buddhist masters, including fragments of texts by Lerab Lingpa and Düdjom Lingpa (Chapter 6). Centuries of practice by innumerable contemplatives have been testimony to, and provided evidence of the rich understanding of the nature of the psyche and the mind, that meditation practices, including Settling the mind, open up to. Some bit of first person experience has been added in Chapter 5.

Viewing this phenomenological meditation approach as one important methodology among other methodologies in getting to a deeper understanding of the psyche is crucial. Both the phenomenological approach – as exemplified and brought to great heights in meditation – and a more structural-developmental approach – as understood in developmental psychology – have much to teach here. The specific practice Settling the mind in its natural state, with observing as barely as possible, with the initial double focus of space of mind and what arises, and the release of grasping to thought without banishing the thought, has its special contribution. As we saw, it combines the power of shamatha concentration-calm meditation with some insight in psyche dynamics (Chapters 8, 9). With deepening practice, there can be an optimally functioning person with a pliant and supple mind (relative), with True Nature (absolute) shining through (Chapter 4).

Settling the mind meditation offers a rich foundation for subsequent mind and awareness practices in the traditions of Mahamudra and Dzogchen (Chapter 7). From these experiential remarks, as will be corroborated with more evidence below, it may be concluded that the practice Settling the mind in its natural state does add to our phenomenological understanding of dynamics in the psyche.

11.2 Wholesome effects for every human being, irrespective of worldview

Hypothesis 2: "Shamatha practices, and specifically Settling the mind in its natural state, in a relational-compassionate context, can have wholesome effects for every human being, irrespective of worldview. This is the case, because, also when stripped of explicitly Buddhist context, this practice can be presented as an attention and emotion regulation strategy and training in its own right. As the practitioner progresses in the training, worldview-related thoughts will less and less be grasped at and identified with."

Specifically, I will address two aspects. First is: the relationally embedded practices having wholesome effects. Second regards the notion that worldview-related thoughts will less and less be grasped at and identified with. This may be seen as one example of a shift to increasing mindfulness and empathy, and part of dynamics of "state to trait," that will also get attention here.

Relationally embedded practices

Regarding this first aspect: in Chapter 4 the need for practicing the attention strategies together with the Four Qualities of the Heart has been emphasized. In more general terms: Daniel Goleman, psychologist and world-wide bestseller writer, advances in the foreword to Alan Wallace's *The Attention Revolution* that this approach offers a potential cure for the chronic distractibility that has become the norm in modern life. What is achieved with attention practices, according to Wallace, is a strengthening of one's psychological immune system, and a heightened sanity. This will refer to the personal and interpersonal aspect: the immune system, ready to receive unwished for intruders, takes their power away and keeps the receiver resilient. We may think of white blood cells that take away the power of bacteria. As in our bloodstream, intruders may show up in our own mind stream.

These shamatha practices do not demand allegiance to any religion or philosophy, and they have indeed been embedded in various secular programs, that teach attention and emotion regulation. One example is the Cultivating Emotional Balance project, named in Chapter 10.

Worldview-related thoughts, and the question: can our states alter our traits?

Regarding the second aspect: as the practitioner progresses, there is more pacification of the mind, more joy, more physical and mental ease. With more meta-awareness, one is less bound to conceptual thinking. While these aspects have been described by many Buddhist authors, for me in first person experience, less grasping indeed included a gradual non-identification, with decreasing self-attachment. This hints at a loosening of self-definition within unquestioned worldviews (for a diary example, see Chapter 5 nr 30: a sense of freedom of not being bound by beliefs, convictions, expectations).

State to trait dynamics and affirmative research for meditation have been briefly explored in Chapters 8 and 10, in what may be called first person Experiences, and third person Behaviors perspectives. Rick Hanson, neuropsychologist and Buddhist teacher, clarifies the mechanism: "Because 'neurons that fire together, wire together,' momentary *states* become enduring *traits*. These traits then become the causes of more wholesome states, which nourish your traits further in a positive cycle." Neurons firing together, wiring together: referring to the famous one-liner by psychologist Donald Hebb. As to firing together: combining and doing things more often creates stronger connections in neuronal networks. [1] Trait formation in this context coincides with creating more and new wholesome habits.

The scientific research in the Shamatha Project addresses state and trait aspects, in the research question: "Do the changes (like: improvements in attention, maintaining resilience in the face of stress) persist after meditation trainees return from the retreat experience to the cacophony of everyday life ...?" (Chapter 1). The pre and post assessments for us participants include a state-trait anxiety inventory: is there a difference in how we can handle anxiety? Besides positive first person calibration, anecdotal evidence, second person intersubjective impressions, also rigorous third person findings are now pouring out. These show an increased ability to sustain attention, less impulsive reactivity and more healthy emotion regulation, in the participants after retreat. And there was this unique finding of the heightened telomerase levels, this connection of meditation and positive psychological change with changes on the DNA level; participants with the highest increase in telomerase being the ones with the greatest improvement in some of the psychological tests. It was found that high telomerase activity was due to the beneficial effects of meditation on perceived control and neuroticism, which by themselves were due to changes in self-reported mindfulness and sense of purpose (Chapters 1, 10). May we speak of supporting findings on the DNA-level, connecting states with traits?

It sounds so logical: with more relaxation, and less self-grasping, there will be less self-judgment and low self-esteem, and *more trust*; and as there is more space, it's just natural that we see the Qualities of the Heart increasingly flowering, both to ourselves and to others. Summarizing, as to the hypothesis: yes, this practice can have wholesome effects for every human being, irrespective of worldview; yes, states into traits!

11.3 Coping with turbulence, signs of progress, signals for worry

Hypothesis 3: "The meditation practice Settling the mind in its natural state invites for inner turbulence. At the same time, it presents empowering ways of making sense of attentional and emotional turbulence experiences and challenges. Moreover, the practice offers tools for coping with these challenges through enhancing the practitioner's capacity to focus and refine attention, to not be distracted by thoughts and emotions that come up, and by enhancing the capacity to not grasp at, and identify with upcoming thoughts and emotions. These are tools that nourish transformation in a wholesome direction. This meditation may be adapted for various applications that will relieve suffering."

Some remarks are made about two aspects. First regards: "Making sense of," second addresses offering tools for coping with attentional and emotional challenges, while addressing state and trait. Possible applications, including in mental health care, are addressed in 11.6.

Making sense of attentional and emotional turbulence experiences

The Tibetan Buddhist tradition has been shown to give detailed information for making sense of attentional and emotional turbulence. The fifth and sixth stages on the path of meditative concentration carry their names with honor: like *Tamed attention*, and *Pacified attention*. These phases on the path may be rough, but certainly also earlier phases provide glimpses of these dynamics. Explanations as given by Wallace and Sopa have been presented (Chapter 4). According to Düdjom Lingpa, intense pain, paranoia and fear can come up, that may be seen as signs of progress. It has been my first person experience that the non-distraction and non-grasping have provided the stage for a "falling out of habitually conditioned personality," bringing out turbulence. At the same time, these instruments also gave me the opportunity to see turbulence dissolve. For both I've felt like a witness in wonderland! (Chapters 5 and 9). As I experienced and described, in my view, (micro)turbulences could basically either evolve into increasing relaxation and opening, or into contraction and closing. The sense of dissolution of the separate-self matrix will always lead to transformation, with a potential for greater health. Knowledgeable and compassionate support is crucial.

Tools for coping, with the question: can our traits carry our states?

It has been my experience, that the trait-aspect of resilience can be very much strengthened with the Settling the mind practice, as the practice includes the training of the tools for observing, tolerating anxieties, seeing mental objects for what they are, without needing to grasp to and identify with them. With numerous repetitions of that, traits evolve that can carry the turbulent states, and chances are better for a more spacious coping style, and for "signs of progress." A meditator with sufficient resilience for going through the temporary turbulence, with the support of teacher and community, can experience a deconditioning of inner grasping reflexes, resulting in greater health. In case that emergence shifts into emergency, meaning that traits in that moment cannot carry the very intense states, professional consultation is required. Research regarding focused attention has shown how practices of attention, emotion regulation and non-reactive monitoring lead to a decrease in emotionally reactive behaviors. Shamatha Project research supports these findings, as we saw. Conclusion: the hypothesis has been broadly affirmed.

11.4 Consciousness and collaboration between contemplatives and scientists

Hypothesis 4: "Shamatha practices, and specifically Settling the mind in its natural state, offer ways of directly experiencing, and becoming more familiar with various dimensions of psyche and consciousness, with gradual refinement in perception. This creates perspectives for collaboration in consciousness research between researchers in various fields, like for scientists and contemplatives."

I address two aspects. Firstly, experiencing dimensions and states of consciousness, with gradual refinement. The second aspect regards collaboration between scientists and contemplatives. Some remarks about a possible new stance in contemplative research are added.

Refining perception and states of consciousness

Description of various consciousness dimensions in a Tibetan Buddhist view have been given especially in Chapters 4 and 7, specifically for the Settling the mind practice in the context of Mahamudra and Dzogchen. The practice guidance we received during the project retreat has nourished a familiarity with various consciousness states (Chapters 2 and 3). In the description of my meditation experiences (Chapter 5), I refer to different dimensions, and I note the increasing refinement in perception and sensitivity. During this expedition, most clearly the dimensions have been of the body, of the mind, of awareness and of qualities of the heart. There has been a process of familiarizing, and habituating with the nature of the psyche and mind, from coarse to subtle and more.

Scientists and contemplatives

Much meditation research is being done, with increasingly sophisticated methods and broader scope. New avenues of study and research will be developed (Chapter 10). There is a growing recognition that the capacity of long-term practitioners to examine, modulate, and report their experience contributes greatly to neuroscience research. However closely scientists investigate the brain, they will not find a thought, a feeling, a mental event – they find structures, chemicals, electricity, functioning and interacting. And however closely a meditator observes (in shamatha), or explores (in vipashyana) her mind, she doesn't detect a brain mechanism that causes the mental processes she is experiencing. The scientific endeavor can be seen as an ongoing process of wholesome disillusionment; in this we need resilience, and all hands on deck.

The Integral approach is helpful in viewing research projects and directions in a larger context. While most meditation research has been third person approach (objective science), increasingly meditation research is now being done in a way of combining third person with first person report. More perspectives can be added: in fact, I asked a researcher why they hadn't included an anthropologist or ethnomethodologist on the staff; this might have added a refreshing second person view. There have been striking anecdotal happenings (like the one I described as a participant observer: the conflict about shared meditation sessions, Chapter 1.3). Also a systemic view might be added. All these perspectives are contributing their unique view. All in all: there are exciting perspectives in consciousness research and the collaboration of contemplatives and scientists in shamatha and broader meditation research; hypothesis 4 is confirmed.

This is the place to return to dust and trust, in relation to the three steps for good knowledge, as sketched in Chapter 3.6. The steps: if you want to know this, do that: the experiment; then comes the experience, followed by communal confirmation or rejection. They regarded the question: "Does shamatha meditation, and specifically Settling the mind in its natural state practice support the dissolving of dust, and the increase of trust?" The first of these four hypotheses worked out throwing light on the nature of trust and dust; the second led to clarifying how trust and dust relate, the third led me to experiencing how attention for dust opens for trust, and the fourth to seeing how a less dusted mind can be more serviceable in a larger research context. For me the response is a full yes.

An alternative stance for a new amplified science

I would like to add some notes about combining third person and first person research in the approach of *neurophenomenology* that has been mentioned in passing. Neurophenomenology seeks to integrate phenomenology, dynamical systems theory, and experimental brain science. [2] The term was given distinctive understanding by cognitive neuroscientist Francesco Varela, in the nineties, as an approach that draws attention to this need to combine quantitative measures of neural activity with first person descriptions.

Michel Bitbol, researcher in the philosophy of mind and consciousness, with an interest in the influence of quantum physics on philosophy, elaborates on the role neurophenomenology may play. A neurophenomenological stance in research involves a radically new approach of subjectivity, he says: instead of being underrated and neglected, with this stance, statements expressing first person experience are reintegrated into the whole system of discourse. Part of this stance is to avoid the mere juxtapostition of an objective science with a poorly studied subjective realm. Instead, he speaks about crossing the threshold of a new amplified science with its own unprecedented structures, with the appeal: "... do not content yourself with looking for correlations between previously established categories and structures. Instead, show that the very process of interconnection between experiential data and objective data may give rise to new categories and new unexpected structures." Varela's notion of "mutual generative constraints" points toward a process of reciprocal alteration and enrichment of experiential and objective concepts. Bitbol calls for awareness of the ways that objectivity arises from a universally accepted procedure of intersubjective debate. He urges us to see that intersubjectivity should be given the status of a common ground for both phenomenological reports and objective science. He emphasizes the need for amplifying the criteria of intersubjective understanding by refining the stability and sharpness of subjective experience. His appeal presents an inspiring agenda for the future, in terms of practices:

> Think about the most basic presupposition of the process of objectification and of establishment of law-like relations between

> objective quantities: a system of socially regulated *practices*. A *practice* of measurement, calibration, and elimination of noise; a *practice* of feedback loop between experimental activity and tentative theories; above all a *practice* of distancing from subjective connotations of the findings and theories of science, despite their being so crucial as a creative background.[3]

Ah, reading about this in terms of practices I feel warm at heart. Some associations come up, regarding refining the stability and sharpness of subjective experience: practice shamatha, practice Settling the mind! In this context I would like to add: do these practices for cultivating stability and sharpness, needed in studying any kind of meditation; and of course, in a recursive way: for more fruitfully studying these practices themselves! Communicating about first person experiences is supported by the protocol for meditation research, as presented in Chapter 10. Bitbol's plea for giving intersubjectivity the status of a common ground for both phenomenological reports and objective science... it reminds me of Clifford Saron's remarks in the same vein, made in the context of the Shamatha Project research (Chapter 1).

11.5 Concentration-calm and possible practical applications in society

Shamatha practices have much to offer for applications in a societal context. Who will not profit from better concentration and focus! A business woman will, a schoolchild, a gardener, a sustainable environment activist, they all will function better in their pursuits. In describing five attentional abilities in the context of "cognitive rehabilitation," indeed reference was made to the situation most of us start practicing from: first, profiting from some sort of cognitive rehab. This is no luxury, given our common attention deficit and hyperactivity problems. Subsequently we may speak of resilience cultivation and tools training.

As became clear in Chapter 4.5, mindfulness based approaches, with their impressive expansion in many fields of health care and life-style, cultivate both a more broad and less specific set of abilities, in comparison to shamatha. More broad, for instance, in the sense that they include extensive body scan and yoga practices; less specific in that the practices generally include less detailed attention training, and that, most of the time, there is not so much contextualization in a broader framework of an encompassing wisdom path.

In this sense, shamatha may have a complementary offer, in the field specifically with attention focus, emotion regulation and cultivating positive qualities. Settling the mind practice has a special place in the program. I feel privileged in offering workshops in this "shamatha empowerment" field to societal organizations.[4]

In what follows I elaborate on beneficial effects that strengthening attentional-emotional resilience can have, on indication, for people in various situations and conditions.

Shamatha in psychiatry, meditation guidance, and the dying process

1. For people with emotional problems

It has been my experience that there may be a circumscribed group of people suffering emotional and psychiatric problems that can profit, as an auxiliary intervention, from an (adapted) form of mind and awareness practices. For the more basic grounding practices, the word "meditation" may be far-fetched and even confusing. In case persons are interested, have a familiarity already with meditation, or feel naturally inclined to it: structured bodily awareness and grounding exercises, movement practices and breath practices are always a good start.

With shamatha applications, generally the more simple and familiar breath meditations will be the entrance. We in the Project have been practicing the three varieties, of focusing on the tactile sensations of the breath in the whole body, the abdomen, and at the nostrils. In case of turbulence, starting from highest degree of turbulence: most basic will be to help a person to "go in the infirmary" (lying down), or sitting, doing whole body breath meditation, or maybe abdomen focused meditation with body-awareness. Deep relaxing releasing out-breaths are advised, let go ... With moment-to-moment attention: a full-time job. This is for grounding in the body, with anchoring for a turbulent mind. On the basis of how this is experienced, further choices and adaptations can be made. Then, there may be persons who have a specific interest in the mind and like to try Settling the mind. It goes without saying that persons with mental turbulence, who have the feeling this would be like stepping into the lion's den, are strongly disadvised to do so! For those with not too turbulent minds: the crucial training is in observing, detachment and non-identification with mental events. What can be helpful is the aspect of "daring to look" at mental contents, in that sense a *desensitization of anxiety.* A person can feel reassured, in seeing events and general patterns. The same goes for the turmoil, chaos and stimulus load, that may at first feel frightening, but also here anxiety may decrease. Some sort of familiarizing may take place. Here I would just advise doing this practice initially together with the therapist, for not more than a few minutes, as a start, with thorough monitoring.

Regarding persons with psychospiritual emergency: as we saw, in principle, this condition may be seen as a kind of non-pathological developmental crises. If possible, let the processes run their course to completion. As I learned, caution is needed when persons, who have meditation experience in their background, are too eager and expect too much of meditation. Meditation may even have been part of the exacerbation of psychiatric problems.

Clearly, it is of pivotal importance that the practice guidance is professionally informed, very well-attuned in a personal sense, on the cautious side, and offered in a warm, trustful and compassionate atmosphere.

2. For those working with persons with emotional problems

The person who gives the guidance needs to be expert in the practices. Next to that, especially the Settling the mind practice can be of great benefit for psychotherapists and psychiatrists in getting to know their own psyche. They will hopefully have done "learning therapies," in the sense that they got to know the position of being the client, experiencing and observing their personal psyche-dynamics and exploring them in the interaction with their therapist. Settling the mind offers additional and different tools, for getting to know themselves as not-other than their clients and patients. Settling the mind has the interesting aspect, as we saw, that the practice naturally invites for some degree of turmoil. At the same time, one is training the tools to cope with the turmoil, in a way of becoming more present and mature.

Teaching shamatha, including Settling the mind to groups of transpersonal psychiatrists and psychotherapists, has been quite inspiring for me. These were therapists, who had an interest in more subtle and refined states of consciousness. It may be very rewarding to expand the scope of these trainings to other professional groups in mental health care.

3. For those who guide meditators on the path

While Settling the mind practice – with its non-grasping, non-identifying attitude – will be very helpful and valuable for everyone, certainly this will be the case for meditation teachers, and not only those who are on the shamatha path. Also vipassana and Zen teachers meet with practitioners getting into mental turbulence. As with mental health professionals, it is good to have one's own experiences with these meditational turbulences, and it is highly desirable that meditation teachers are familiar with the various states that can be called forth by intensive meditation. A meditator-in-turmoil will feel supported when the teacher speaks and guides from experience. Mental turbulence in one participant in a group tends to ripple into distractions and feelings of insecurity in the group (I remember practitioners asking: "Can this also happen to me?" "Do I need to give more help?"). Here meditation teachers have an important role: they must give realistic explanations, reassure where possible, and support the dedication for continuing the retreat. With a sense of all-inclusiveness: this happens, and it evokes many feelings and thoughts for all involved. And, it's all part of the practice. Sometimes a form of structured support by group-members for the one in need can be helpful. It is necessary that the teacher knows his or her limits in guiding persons who get into conditions of great turmoil, and knows the right moment for getting external expert professional help, as has been addressed in Chapter 9.

4. For persons with attention deficit problems

To the vast number of children and adults, suffering from what are officially recognized as attention deficit problems, most of the time a variety of (combined) treatments and forms of support is offered. For some, problems start with traumatic brain injury. Possible interventions for persons with Attention Deficit Hyperactivity Disorder (ADHD) and Attention Deficit Disorder (ADD) include various kinds of attention training, with psychosocial interventions, medicine, EEG neurofeedback, Heart Coherence biofeedback, classroom programs, and parental involvement. An example of a program that includes shamatha meditation is the Mindful Attention Program (MAP), described by Lidia Zylowska. It is designed to increase attentional focus and the self-regulation of attention through such mind training practices as single-pointed meditation, insight meditation, sensory awareness exercises, and practical strategies to increase awareness in daily activities. In the problems, deficits in executive functioning, including working memory deficits, play their part. Mentioning Heart Coherence biofeedback: as a more general issue, I like to add here that combining biofeedback with shamatha practices proves to be a very fruitful area for application and further exploration.[5]

Diary after retreat, 2010

> Writing about persons with traumatic brain injury ... here again I can take myself as a guinea pig example. Earlier I presented a report about my meditation experiences, after having fallen on my head, on this icy path, during the Project retreat. It returns in conscious felt sense memory: the sound of the backside of a skull on ice ... a bit dry and dull, then silence. Headache, while lying down on this granite-hard surface. And seeing the blue sky! Dizzy, on the arm of a gentle *bodhisattva* to my room. As I wrote, it brought up body-memories and old fear, connected to traumatic brain injury I had in the past, with damage to the fronto-temporal area, front and side of the head. Thinking back of that phase in my life ... As I read: management of attention, working memory and executive functions may be affected. One aspect of executive function is the ability to make mental shifts and engage in flexible thinking. That rings a bell. The inability to do that was exactly what has been quite a confrontation, after this accident. Sometimes it "could be observed," yet, sometimes it just drove me nuts! I recovered ... and then, during project retreat, I realized that it has been Settling the mind that helped me intensifying exercising these shifting-abilities, again and again. Cognitive rehabilitation!

5. As a preparation for dying

Another possible application, very evident for the Tibetan Buddhist practitioner, may seem less evident for the Westerner: the shamatha practices may very well be cultivated as a preparation for the dying process. Especially the Settling the mind practice may be beneficial, both for those who live and work with dying persons, as well as for a person who is ready to face her or his impermanence, and intent on being well aware during the process of dying, when death is near. As has been addressed by Alan Wallace during the Shamatha Project retreat, we are all said to experience substrate consciousness (storehouse consciousness) in dreamless sleep very often; then, lacking clarity. Also, it is experienced in the blackout phase of the dying process, when all sensory and mental appearances vanish. A way of exploring this dimension of consciousness during our lifetime is through the practice of shamatha.

The shamatha practices support one in being more familiar with the spaces of awareness that may present themselves on the journey. While being on this journey all through our lives, we could conceive of doing shamatha meditations, as described, in the last phases in life. First, doing breathing meditation. Then, when the body slips away, when physical senses are said to dissolve into the mental sense: second, shifting to practicing Settling the mind. And after that, when the mind fades, and dissolves into awareness: to third, continue in practicing Awareness of awareness. Dying, with a supple mind, with a psyche dissolving in substrate consciousness, then in the larger space, in awareness.

Contemplating the sequence in this context has added meaning to the shifting with the three practices for me. I feel supported by this approach, and have shared it with colleagues with whom I'm working in hospice care. I've been fortunate to meet Christine Longaker and Joan Halifax Roshi who have done such pioneering work in this field, and feel inspired with their writings. In my experience, some familiarity with adapted forms of shamatha attention meditations and the Four Qualities of the Heart, complemented with *Tonglen*, and *Phowa* can be so helpful for hospice workers. *Phowa,* a Vajrayana practice often referred to as "transference of consciousness at the time of death," can also favorably be done at any time during life. It is about uniting your mind with the wisdom-mind of the Buddha, a Sacred Presence, Light. In fact, all of the named practices are meaningful during living and dying, death being an intrinsic part of life. Joan Halifax Roshi, actively engaged Buddhist in many fields in society, considers the great gift of the Four Qualities of the Heart that they can live in the background of our mind and heart: they are always there, when we lose balance, in righting ourselves, in facing suffering, our own and the suffering of others. These practices are valuable for both the dying and the caregiver. We can choose and compose phrases that are personally meaningful. May Loving Kindness fill and heal your body. May I take care of myself. May joy fill and sustain you. May you accept things as they are.[6]

Compassion and dying – guided meditation

Let's do a guided practice, in this context. The complete meditation, presented by me during a shamatha retreat, includes compassion and *Tonglen* practice, addressing impermanence, sickness, old age, and death. Here I present the part on dying. It feels right to tell you in advance that this meditation will include imagining your own dying process. If you are not ready for that, graciously, with compassion for yourself, skip this meditation. Maybe do it at a later moment, a moment that feels right for you.

> Let's settle body, speech and mind.
> Now, bring to mind *a person who is in the process of dying* – maybe a person you know, or a person you knew in the past, and that you cared for when the person was dying; or a person that you know, who is soon going to die. Like we, at one time, are all going to die. Let the mind be in its creative, spacious dimension that transcends self-attachment: your true nature, symbolized as an orb of white light at your heart. Imagine yourself in this person's position, in the felt sense: maybe lying down in a bed, feeling very weak and exhausted, hardly breathing, or gasping for air; maybe in pain, maybe with anxiety, feeling lonely. Possibly, there is a sense of balance, surrender. Bring in the details for all the senses, as you imagine these, when being this person: what you would see, hear, feel, taste, and smell, and what your thoughts and emotions might be like. Whatever they are, be present with them, with compassion for the person, and for yourself imagining being the person.
>
> And now, step back into your own perspective. Breathing in,

the possible unrest, the fear of this person; let them dissolve, without a trace, in the light at your heart. Breathing out, send love and compassion, caring and warmth. And let go ...

Now, imagine *your own dying process*. We have no idea about when and how this will happen, how long it will take, how aware we may be when the time has come. We may wonder: will I be able to be there with presence, wakefulness? Imagine yourself dying, and feeling bodily sensations getting more thin, more fleeting, more light, maybe more dull ... and be with that, keep doing breathing practice; whole body, abdomen, nostrils as the focus of attention. Imagine that your breathing may become so shallow that you can hardly recognize it, that it fades away. Still there is mental activity, you can be aware of thoughts coming up and dissolving, gradually fewer, you may be aware of the space of mind that more and more is just that; a space. In a way, you are practicing Settling the mind in its natural state ... Then all mind fades, and there may be Awareness of awareness ... and then, just awareness. As for the other person, so for you, so for all of us, when the time has come ...

And now, let go of all images and imaginings. Step back into your perspective as it is, now. Sense your arms and legs, feel how you sit, be aware of looking, hearing, of where you are, here, now. Be your aliveness! Live it, and celebrate. And know that these practices are always with you, in living and dying; that this spacious awareness is always with you, that this lively awareness is who you are.

May this meditation be of benefit, not only to ourselves but also to all those around us.

Also for you, reading these pages: sense your arms and legs, feel how you sit, be aware of looking, hearing. Maybe, stand up and walk a bit. Be here, now, be your aliveness!

An undivided love

While writing this, I hear a song in my head. Some lines in "Come Healing," by Leonard Cohen, singer and Zen practitioner, arise:

"O troubled dust concealing an undivided love
The Heart beneath is teaching to the broken Heart above."[7]

I don't know why these come up now. It's about compassion and love, for the other person, for the human situation, for ourselves, for all sentient beings. It's about undivided love, and it's about dust. The privilege of being with a dying person, the sharing of these precious deep, raw, existential moments, includes the invitation for bareness, presence with not-knowing, and facing whatever dust is concealing this undivided love – or, suddenly, lifts. The Heart beneath: the love that is our nature, always present, for our broken Hearts above.

11.6 In a larger evolutionary context

With shamatha, we may get inklings of the qualities of bliss, luminosity and nonconceptuality, ascribed to attaining substrate consciousness. We may experience this undercurrent of joy, love, compassion. Interestingly, some people, who claim to have had a near-death experience (NDE), have talked and written about what they went through, after "coming back." This coming back is felt as from (almost) death. Connecting with Wallace's remarks: it may be seen as "from substrate consciousness experience," and probably even more subtle. Persons after this experience often talk about giving greater value and meaning to life and less importance to material things. Pim van Lommel, Dutch cardiologist, who gained world renown for his accurate controlled study with persons with cardiac arrest, connected with their heart attacks, reports on their experiences of near-death. He notes: "The near-death experience turns out to be a life-insight experience. Or as one person who experienced a NDE put it: NDE really stands for New Discernment through Experience." And he adds: "The newfound insight pertains to what matters in everyday life: acceptance of and unconditional love for oneself (including acceptance of one's dark side), others, and nature. It also pertains to insight into connectedness: everything and everybody is connected." Anita Moorjani, after the NDE she went through, wrote the following words in relation to love and connectedness: "Each of us, at our core, already is pure and unconditional love ... It starts by loving and trusting myself. The more I'm able to do so, the more centered I feel in the cosmic tapestry. The more connected each of us feels, the more we're able to touch others, enabling them to feel the same."[8] Trust! From unexpected sides, quite beautiful references to experiences and meanings where also shamatha practice, including Settling the mind practice, and deeper insight practice, may bring a person. Imagine, continuing on this line of: where shamatha practice may bring a person. It may not only be one person: many persons make a society. Imagine more people having these life-insight experiences, the experiences of connectedness, not after NDE (in this official conceptualization of massive confrontation with mortality and impermanence), yet, with shamatha practice. Any measure of shamatha accomplishment brings us increasingly in touch with these qualities and experiences.

The importance, and at the same time for Western scientists, the mystery of substrate consciousness ... This notion has also for me been so crucial, in practice and understanding. Thrangu Rinpoche has been referring to it as all-basis-consciousness, the basis for the arising of all other types of consciousness, the ground for the sensory consciousnesses like seeing, hearing, and the mental consciousness; constantly present. Alan Wallace, with a scientific flavor, states that substrate consciousness may be characterized as the relative ground state of the individual mind; it entails *the lowest state of activity, with the highest potential and degree of freedom* that can be achieved by evacuating the mind by way of shamatha practice. I feel driven to transmit some of Alan Wallace's far-reaching musings and positions, relevant in this context of Buddhist wisdom and Western science. In his view, with an analogy from modern biology, substrate consciousness may be portrayed as a kind of *stem consciousness.* In a way like a stem cell differentiates itself in relation to specific biochemical environments, such as a brain or a stomach, the substrate consciousness may be said to differentiate with respect to a specific species. As Wallace advances: for the Western materialistic scientist, the existence of substrate consciousness belongs to the realm of metaphysics. For the contemplative adept, however, with thousands of hours of continuous training in developing mental and physical relaxation, together with attentional stability and vividness, it is an empirical fact. A fact, assessed with this highly refined attention, like a telescope, directed to the inner space of the mind.[9]

Alan Wallace compares the substrate consciousness to a "relative vacuum," relatively empty, but still possessing structure and energy, characterized by the named attributes bliss, luminosity and nonconceptuality, with a muted sense of duality between subject and object. Primordial consciousness, the absolute ground, could be characterized as the "absolute vacuum" of consciousness. He quotes theoretical physicist Henning Genz: "Maybe the true vacuum, the true nothing of philosophy and religion should be seen as a state wholly innocent of laws, space, and time. This state can be thought of as nothing but a collection of possibilities of what might be." According to Wallace, at the level of primordial consciousness, no distinctions between matter-energy and consciousness are found. They are undifferentiated, yet, they are accessible to the human mind. "Whereas the inferences made by physicists speculating on the vacuum are conceptual and lacking direct experience, contemplatives with the ability to rest in the open clarity of primordial consciousness may perhaps experience the source of universal creation. From a common-sense point of view, that ... would seem nothing less than miraculous." With these breath-taking perspectives: next to measurements, and assessments of brain correlates to mind states, there may be research on notions in Buddhism that are unfamiliar to, and not known in Western cultures. What might Western psychology and consciousness studies learn from the notion of substrate consciousness, if it was taken as a working hypothesis, to be tested? About this ground state of the mind Wallace expresses his belief that this is one of the most remarkable discoveries ever made concerning the nature of consciousness. For him, this ground state, this natural state, calls for collaborative research between cognitive scientists and contemplatives. [10] So, here's another plea for intensifying collaboration between researchers, scientists and contemplatives, in consciousness research, with wide perspectives for investigation breaking fresh ground.

Chapter 12

On human flourishing: continuing the project

> *The mind is constantly in motion, but in the midst of the movements of thoughts and images there is a still space of awareness in which you can rest in the present moment, without being jerked around through space and time by the contents of your mind.* (Alan Wallace, 12.4)

What does it mean, settling the mind in its natural state? The least thing I can do is to bring a number of fresh notebooks and pens for jotting down, reporting. These words I find in my diary, in a fragment of one of the first days of my participation in the Shamatha Project retreat, in September 2007. The words are also presented in the beginning of what was to become a book. During retreat, I could imagine trying to further explore this question, also possibly after retreat, by meditating, doing some reading and writing. I couldn't have dreamt that the question would lead to *this* book. Ongoing involvement with the mind has made me very much aware, respectful, humble, and in awe, as to what we make of the mind, what the mind makes of us, as to what is the nature of mind, to how we are no other than mind.

This last chapter has two main themes. The first includes some musings on human flourishing. It is about practicing and being "in the world," but not "of the world," it's about a natural state, and it's about "doing what needs to be done." In a broader sense, this is about continuing the project of our shared humanity, and about flourishing for all sentient beings on the planet. The second theme regards, literally: continuing The Project, in the sense of the Shamatha Project. Recent funds have been secured, facilitating the continuing of the Shamatha Project, with a new round of follow-up measurements lying ahead.

12.1 The most compassionate thing we can do

Connecting with flourishing: the one meditation that we did during the Shamatha Project that has not yet been shared with you in this book is Empathetic Joy.

Empathetic Joy – guided meditation

Following, for starting this chapter, and rounding up this book: find a practice guidance, inspired by some themes Alan Wallace addressed during retreat and also in his book *The Attention Revolution.*

> Let us find a comfortable position. We keep the spine straight and not rigid. Settle your body in its rest state, imbued with the three qualities of relaxation, stillness, and vigilance. Relax and release. Now, bring to mind a person you know well who exudes a sense of good cheer and well-being. With all your senses, imagine being with this person. Think of this person's physical presence, words, and actions. While attending to this person's joy, open your heart to that joy.
>
> Now bring to mind another individual. Think of a person for whom something wonderful has happened, it may be recently or in the past. Recall the delight of this person and share in the joy.
>
> And now, direct your attention to someone who inspires you with his or her virtues, such as generosity, openness, kindness, and wisdom. Rejoice in these virtues for this person's sake, for your own sake, and for all those who are recipients of this virtue.
>
> Now direct your awareness to your own life ... Attend to periods in your life that have been inspiring and enriching to you and perhaps to others as well. Think of occasions when you embodied your own ideals. This is about your virtues: attend and take delight in these wholesome qualities in yourself. There doesn't need to be any pompousness here, or any sense of pride or arrogance. As you remember the people and circumstances that enabled you to live well and enjoy the sweet fruits of your

> efforts, you may simultaneously experience a deep sense of gratitude and joy. This prevents you from slipping into a superficial sense of self-congratulation and feeling of superiority. Take delight in these memories and the virtues they relate with.
>
> Some practices are difficult, but the practice of empathetic joy is easy. When throughout the course of the day you see or hear about someone's virtue or good fortune, empathetically take joy in it. It will raise your own spirits and help you climb out of occasional emotional sinkholes of depression and low self-esteem, that may occur during meditation. [1]

These were the kind of instructions Alan gave us, during the project, in a situation that included a relative isolation for us, where we didn't talk much. Evidently, quite some surprises might come up, including boredom and restlessness, and for some of us depressive feelings of self-doubt: it is in that context that the last lines in this guidance for Empathetic Joy are phrased. Next to some possible small-perspective personal fixations and worries, there can be Empathetic Joy, for ourselves and others. Empathy means: feeling "in" others, so to say, in the sense of the ability to share someone else's feelings or experiences by imagining what it would be like to be in that person's situation. This practice inspires, not only for moments of feeling low, but for any moment. Taking delight in others' and our own virtues, while being an important ability, is often overlooked.

With Loving Kindness and Compassion meditation, we cultivate the deep wish that others and ourselves may find happiness and its causes and be free of suffering and its causes. Next to that, the cultivation of Empathetic Joy attends to something that is already a reality: the joys, virtues and successes of ourselves and others.

I have felt privileged with a deeply dedicated teacher, presenting shamatha attention and Heart practices in a way of emphasizing their many benefits for daily life. In a general sense, I have learned that the shamatha meditations can, after initial instruction, be practiced by a person who has no direct connection with a teacher. Regarding the place of shamatha, during the project, the meditations were presented as possibly standing by themselves, with their specific merits, and also as the foundation for further-going insight practices. Alan Wallace has taught us thoroughly in the importance of shamatha, that may be practiced "all the way" along the Elephant Path. In addition, it is possible to apply these meditations more directly as preparatory for insight and Essential practices. In the Essential context shamatha may be practiced in a way, sufficient to cultivate the "staying" power and pliancy of the mind as considered appropriate. All: flourishing ...

A spirit of emergence, and nadi-plasticity

Happy and unhappy person, happy and unhappy mind ... Almost anything can catalyze unhappiness, but the true source of either happiness or unhappiness is always in the mind. I was able to observe this, time and again, in practicing Settling the mind in its natural state. We feel unhappy as our minds are so often unbalanced, obscured, with dust on our eyes, and in our mental functioning. With dust on the eyes, rust on the nerve-endings, sand in the machinery we can't open for, perceive, be with, "be" reality. Which leads us to seek happiness in the wrong places.

I like this sad expression of the "hedonic treadmill," coined by psychologist-neuroscientist Richard Davidson, in relation to ways that persons usually seek happiness, in the conventional, "hedonic" sense. We can be caught in it, always seeking more material things, more stimuli, more enjoyments. [2] It has an addicting quality to it. Disappointment and desperation often seem to reinforce the tenaciousness and, so to say, hidden "narcissistic rage" behind the continuation of this search, outside of ourselves.

And then: there is the expression of the "spirit of emergence," with a shift of priorities. We move away from the sources of discontent and set out on the path to genuine happiness. As Alan Wallace advances: this is the most compassionate thing we can do for ourselves. Like one participant noted, in the questionnaire after a shamatha retreat: "I think being in retreat has been the best time ever spent in my life and the most deep act of love with myself." These are the wonders of the mind: with our mind we can change our relationship with the mind, to the arisings in the mind, and to the mind dynamics. The mind, the relating, and we, change in the process. There can be a shifting in vantage point, time and again, moment by moment, correlating with neuron by neuron, cell by cell. We have to be, it seems, many times in a state of misery, or be confronted with endless repetitions in our craving states and habits ... and then, when awareness of space, of alternatives and choices opens, suddenly there can be this shift in balance: it has been enough, enough times of abiding in these states of unease. To me, it connects with one of the findings in the Shamatha Project research: the increase in the *sense of purpose* in life, in the participants. In their article, Shamatha Project researcher Tonya Jacobs and her group members describe how this change may occur, "as intention and priorities shift away from hedonic pleasure to more genuine contentment and a greater sense of contributing to human welfare."[3] Something seems to settle in our minds: to settle, in a more natural state. In the Buddhist view, a meaningful life with purpose connects with three kinds of pursuit. These are: first, the pursuit of ethics, relating to values, beliefs, and virtues. It is also about the choices we make: how we want to live together, in what kind of a world. Second is the pursuit of mental balance, with concentration-calm. Third is the pursuit of wisdom. This includes insight into who we are, what is our relation to others and the world around. All three are connected with an authentic aspiration, oriented toward, and leading to genuine happiness. The three pursuits are deeply related, they have been named in traditional Buddhism, and they are ever fresh, also in our contemporary world.

Overall, the good news is that we can modulate and un-learn unwholesome patterns. And, at the same time: by cultivating certain practices and abilities we can learn in wholesome directions. This has always been known by those who studied and trained in these practices, in an experiential sense. While not carrying that name, and without a notion of how to imagine it: there always has been neuroplasticity. Now, some people who just believe their eyes when they check phenomenological realities also by way of neurophysiological correlates in the brain, can also be witnesses. "Neurons that fire together, wire together": as combining and doing things more often leads to stronger connections in neuronal networks, every experience, thought, phantasy, counts. As my experience, and my attitude toward my experience, change, the make-up of my brain changes. As to embodiment, there's not only neuroplasticity, the circuitry is not only in the brain. It's all-over-plasticity and dynamism, in every sentient being, in all the billions-trillions of cells with their cell divisions and chromosomes. Here as well: on the DNA level, and beyond ... Talking in terms of the Tibetan inner yogas and the subtle body, this is about the very substrate of our bodies: we may call this "*nadi*-plasticity," with a term used by meditation teacher Andrew Holecek.[4]

12.2 A new round of follow-up for the Shamatha Project, six years later

For this concluding chapter, here is a brief summary of the main outcomes of the project research. Talking in terms of "we" for our group of research subjects, it sounds something like this: it has been found that we participants reported improved psychological functioning, we were better able to sustain visual attention and inhibit habitual responses. Next to that we were more engaged with and sympathetic to suffering. We showed greater activation of attention-related brain regions during meditation after training. Additionally, we had improved measures of cellular health that have been connected to aging. Those of us who reported greater mindfulness had diminished stress hormone levels. Thank you, researchers!

The news came in fall 2012. This is the headline message of the news-release: The Shamatha Project has been awarded a grant of 2.3 million dollar over three years to continue and extend this most comprehensive investigation yet conducted into the effects of intensive meditation training on mind and body. The Grant, titled "Quantifiable Constituents of Spiritual Growth" will support the latest phase of research that will address two questions. Question one: "After going through intensive meditation training, what differentiates people who develop their lives in ways that relieve suffering for themselves and others close to them from those who do not?" And question two: "How are measured changes in cognitive, psychological and physiological processes related to people's life experience years later?"

As the news release announces: next to aiding in the completion of analysis of the original data set, the new funds will support follow-up data collection. In the new round, the researchers will carry out structured telephone interviews with the participants, assessing their experiences of the retreat six years later. They will investigate what changes retreat experience made in the participants' lives, and how those changes continue to affect the participants. Clifford Saron, Director of Science, Baljinder Sahdra, Co-Director, and colleagues will use a sophisticated network analysis to see which physiological and psychological measures made during the retreats are associated with long-term personal growth, after these years, and which are not. Additionally, the researchers will interview family members, colleagues and friends of the retreat participants, to garner their observations about the long-term changes in the participants.

This is wonderful news, and, with the scientists, we participants feel excited! Certainly, the last line about including family members and colleagues, our dear and near ones, gives me a thrill, and touches my heart. This interpersonal "conduct" in society is where the fruits can lie! Indeed, Saron refers to this aspect: "We view it as a commitment to investigate the nature of one's mind in a developmental process of becoming familiar with 'the world within.' This promotes a more knowing and friendly attitude toward oneself. We think this greater comfort 'within our own skin' will be reflected in mental and physical health, our actions in the world, and felt by those with whom we interact." Saron notes that the sense of "purpose in life" is gaining increased recognition within the field of psychology. This aspect, also included in our Shamatha Project research, is seen as a key to sustained health. In Saron's view, the project has a wide potential impact to health and medicine, law, business and society at large. He has spoken about the Shamatha Project to very diverse audiences, including prison administrators, major corporations' staff and agricultural leaders.[5]

International Shamatha Project

The Shamatha Project, with its scope and vision, is expanding in another way. Since our Shamatha Project retreats in 2007, there has been the birth of the International Shamatha Project, initiated by Alan Wallace. It is modeled after the Human Genome Project in the sense that it entails the collaboration of many scientific laboratories throughout the world. The International Shamatha Project will bring together dedicated Buddhist teachers and meditators, collaborating with psychologists and neuroscientists. Their research with shamatha practitioners includes investigating which methods of shamatha are most appropriate for which kinds of people in the modern world. The Dalai Lama, in his written endorsement to the Project in 2009, shows himself a warm supporter. [6]

12.3 "Settling" the "mind" in its "natural state"

In the course of the Project retreat, next to wondering about the meaning of the "natural state of mind," I've been intrigued by the meaning of the other words in this translation of the practice's name. About *settling*, felt-sense: what kind of a verb or noun is this settling? Is there anyone who is settling anything? Is there any subject and object, is there an action "settling," or noun, "the settling"? In a sense, in the beginning of Settling the mind shamatha practice, aiming at controlling and taming the mind, there is someone who tries to "do doing this," "someone who tries to settle a mind." Later ... it's the mind, including the settling, non-doing, just being Settling ...

Mind: a discernment of "relative" mind, and mind in "absolute" sense, has been made. As Thrangu Rinpoche clarifies, in the Mahamudra context: there are two aspects of mind, mind as it appears, and mind as it really is, True Nature. It connects with relative reality, on the one hand, and what is referred to as absolute, genuine reality, on the other. The vantage point from which we view the world, and the way we relate to phenomena, is crucial.

Then, connecting to the *natural state*: what is natural? Here, as well, what is considered natural differs depending on the vantage point from which you describe. The natural state may be seen to coincide with various states of consciousness. Two quotes that have been presented in earlier chapters make this clear.

The natural state, as relating to absolute, to *True Nature mind,* pristine, primordial awareness: " ... if you have some recognition of your mind's nature, then, when any one of those thoughts arises, you will experience the mind's true nature in that thought, because the mind's nature is also the nature of that thought," says Thrangu Rinpoche, in the Mahamudra context.

The natural state, as relating to relative, *substrate consciousness* – Padmasambhava's words are: "settle the mind in its natural state by letting it be just as it is, steadily, clearly, and lucidly ... in the mind's own mode of existence." This is the reference to quiescence in Dzogchen context, and this is also the way that Alan Wallace refers to the natural state in relation with substrate consciousness, mind as relative mind.[7]

And, relating to relative as *habitual psyche-consciousness,* in the gross sense: some may refer to this as a natural state, a state that is conventionally seen as the state of the so-called healthy person, in the sense of absence of gross illness and pathology. This natural state of mind probably refers to a mind with the average, habitual degree of thoughts, emotions, attachments, and dust. This is what, in Buddhist psychology terms, we would consider a mind bound to grasping, an unhealthy mind. Much more real health is possible.

12.4 Doing what needs to be done, just being

Less dust, more trust, in an increasingly natural state, from increasingly subtle vantage points: relating to psyche, to substrate consciousness, to primordial ... with primordial awareness always shining through. All potentially accessible at any moment. Transcending and including ... Less dust manifests in decreasing defensiveness, in ego-personality "getting out of the way." It manifests in decreasing "doing," more being and non-doing, in the sense of a natural "doing what needs to be done," that comes from a deeper knowing.

It continues to intrigue me how, for Settling the mind, the turmoil in the mental and emotional sense is observed with seeing events and patterns occurring "as if under a transparent bandage." It seems to have an impact, on daily life after retreat.

Diary after retreat, 2012

> I'm aware that our three months of shamatha practice is peanuts in comparison with monks and nuns who have done these practices for thousands of hours. Correspondingly, the effects in us, post-shamatha retreat, and coinciding research outcomes are of quite different caliber. Still, I'm amazed what this relatively short period could do. For instance, for me in a sense of inner transparency, and in accessing this inner "still place," in Settling the mind.

"The union of stillness and motion": these words that Alan Wallace presented in a meditation guidance come up again. "The mind is constantly in motion, but in the midst of the movements of thoughts and images there is a still space of awareness in which you can rest in the present moment, without being jerked around through space and time by the contents of your mind. This is the union of stillness and motion" (Chapter 9.2). I'm aware that for me "stillness and motion" thoroughly connect with both the shamatha attention practices and the Four Qualities of the Heart.

Regarding our relationships and the world, in connection with shamatha and Qualities of the Heart: increasing inner peace has had an effect, I realize, on my commitment to outer peace: in relationships with others, in the world, for a sustainable planet. Inner peace, directly connected with "outer peace," requires an inner meditative "activism." Outer activism can be the more effective when it's done by a person abiding in inner peace. I'm reminded of the exchange we had with Alan Wallace, during the Shamatha Project retreat, described in "*Tonglen* and state terror" (Chapter 3). We can practice *Tonglen* for those who suffer oppression. We can meditate, and prepare the body-speech-mind for the right action, for the moment that the opportunity offers itself; and we can go out in activism. Settling the mind meditation, I think, is a practice that supports us in preparing ourselves to "do what needs to be done," also in the outer world, without getting polarized inside ourselves, or becoming part of polarizing between groups.

Four aspects

I want to finish with naming four aspects in this quest that have touched me in a very personal way, regarding this exploration in meditation, science and psychology. I refer to them as:

1. Settling the mind in its natural state, a gift

While we direct our attention to the space of the mind and the events that arise, with non-distraction and non-identification, grasping to a thought is released, while the thought is observed, and not banished. While there is a witnessing of the turmoil in mental space, there is the training of the tools for coping with the turmoil, and becoming more stable. With softening self-attachment, the practice leads to greater peace of mind, getting out of reactivity, and as an outflow, getting to be of more help to others. It remains a practice that I do on a regular basis, together with breathing-abdomen shamatha, as foundational for practicing Mahamudra and Dzogchen. My regular meditation package also includes one of the Heart Qualities, and Kum Nye Tibetan yoga, with walking meditation.

2. *Access to all states, spaciousness*

This aspect regards the realization that any person can potentially have access to all possible states of consciousness, in any moment; including very coarse, up to awakened. This immense range! I gave myself as an example: I've let you share in my diaries with descriptions of coarse states of anger, obsessiveness, paranoia ... up to much more subtle states with experiences of bliss and clarity. This feels like stretching and transcending habitual imaginary self-set limitations. It has deepened trust: firstly, as I get to realize that this vastness and spaciousness is there, and being it. Secondly, experiencing the ability to tolerate the coarseness that transforms in the process. And thirdly, as this all has brought me more in touch with genuine happiness and flourishing.

3. *States to traits, including loosening attachment to self: evolution*

This regards the finding that, with observing and experiencing states of mind, I've trained traits of resilience. Witnessing the natural healing tendency of the mind, melting emotional turbulence and strangled self-attachment, I've experienced how this indeed can feel very therapeutic, and opening: first in the sense of coarse states and hidden materials, and then in the sense of loosening attachment to self. More refined states of awareness are accessible, with greater presence, translating into character traits and more sustainable stages of development, with greater maturity and freedom. In Western parlance of attachment style (different from the Buddhist parlance about attachment and grasping) we may speak of shamatha meditations supporting the developing of an acquired secure attachment style, with greater trust and well-being.

4. Seeing correlates in context, with widening perspectives

This has touched me, in a sense of feeling supported by the Integral approach: seeing the many various perspectives, that can bring in deeper understandings about meditation, consciousness, and the brain. Recognizing how all the various disciplines bring in truths that deserve recognition; truths, in the sense of partial truths. Seeing and being the dance in self and consciousness: I am not my meditation experiences, I am not my brain EEG patterns ... Seeing how, in a transcending research stance, the very process and practice of intersubjectivity can be a common ground for both phenomenological reports and objective science. Seeing how interconnecting between experiential data and objective data may inspire new categories and new unexpected structures. Exploring how findings in the many dynamic domains can contribute to less suffering and greater happiness in the world.

Wish

This book has taken you, reader, on an exploration, on a quest. Next to reading about my quest with shamatha and specifically Settling the mind, may you cherish your own quest, and may the ingredients offered in this book be of benefit.

Notes

Introduction

1. Thanissaro (transl.), 1997; Harvey, 2013: 22.

Chapter 1 The Project, the research, and some outcomes

1 Buddhaghosa, 1999: 85, Wallace, 2005b: 308.

2 Singer, 2010: 171.

3 For an overview of the Shamatha Project, including web links and information about the investigators, funding, and for up to date outcomes, with the complete articles-texts in pdf. format: see Shamatha Project, in the References. Recently a detailed publication on the Project came out, written by Saron, 2013. Overview articles in various journals include: Bond, 2011, Fraser, 2011, Marchant, 2011, and Van Waning, 2011.

4 For some recent works in this field, see for instance Begley, 2007, Lutz et al., 2007, Goleman et al., 2008, Austin, 2009, Hanson and Mendius, 2009, and Siegel, 2007, 2010.

5 Fraser, 2011: 29.

6 Meditation guidance transcripts with permission.

7 Nyoshul Khenpo Rinpoche (1932-1999), see References.

8 Wallace, 2006: XI, Begley, 2007: 215.

9 Association for Psychological Science, News Release, July 14th 2010.

10 Gallagher and Zahavi, 2008: 7; Clifford Saron, personal communication, February 13th 2013.

11 Ekman and Friesen, 1978, Ekman, 2007.

12 See Epel et al., 2009, on meditation and cellular aging.

13 Saron, 2009, Wallace, 2009a: 114.

14 For the Project publications referred to in Chapter 1, see References. Direct quotes in this section 1.5 have been taken from the article summaries, as indicated with page number: see MacLean et al., 2010 (829); Jacobs et al., 2010 (664-5); Sahdra et

al., 2011 (299); Saggar et al., 2012 (1), and Jacobs et al., 2013 (1). An overview of the findings so far is presented by Saron, 2013. For more information and linking to all articles in pdf. format, see Shamatha Project, in the References.

15 Singer, 2010: 174.

Chapter 2 Attention meditations

1 Wallace, 2005a: 26.

2 How could they measure this, so many centuries ago, I wondered. The answer is in the water clock, a timepiece in which time is measured by the regulated flow of liquid into or out of a vessel where the amount is then measured. Allegedly, time clocks have existed many centuries BCE in Eastern countries. Gethin refers to figures, presented in traditional texts, of between 0.13 and 13 milliseconds per pulse. He points at the Buddhist understanding that, when analyzing reality down to the shortest conceivable moment of time, what we still find is a process rather than inert, static bits (Gethin, 1998: 221-2).

3 Wallace, 2006: 136-7.

4 Wallace also gives an overview in a table of the nine stages in his book *The Attention Revolution*, 2006: 174-5.

5 Wallace, 2006: 64, Harvey, 2013: 328-31.

6 Kalama Sutta – To the Kalamas (AN 3: 65, A I 188). In Bodhi (ed., introd.), 2005: 89-90.

7 Anderson and Hopkins, 1991: 154. Regarding Ridhwan: the Diamond Approach is the spiritual teaching, the path, and the method of the Ridhwan School. As a contemporary spiritual teaching, the Diamond Approach to Inner Realization considers the totality of the human being. It is a path of self-realization and human maturity based on an original synthesis of modern discoveries in the field of psychology and a new paradigm about spiritual nature.

8 Harvey, 2013: 50-52.

9 Harding, in Kongtrul, 2002: 3-4.

Chapter 3 Qualities of the Heart and *Tonglen*

1 Walsh, 1999: 99-103; quoting Rumi in Shah (1971: 357), Lau (1979: 116) and Perry (1981: 608).
2 Wallace, 2006: 125, 2009b: 79. See also Tenzin Wangyal Rinpoche, 1998, LaBerge in Wallace (ed.) 2003: 233-258, Wallace, 2012.
3 Brown, 2006: 60-2.
4 Ingersoll and Zeitler, 2010: 19.
5 The three steps: based in Wilber, 2006: 267.
6 Wilber, 2006: 19-23; Esbjörn-Hargens, 2010: 33-61, the full article, including illustrations, can be found on the internet, see References.

Chapter 4 Shamatha in Buddhism as it developed in Tibet

1 Lutz et al., 2007: 509.
2 Harvey, 2013: 414-18; Harding in Kongtrul, 2002: 6.
3 Brown, 2006: 8-9; for naming the Essence traditions, see also Dzogchen Ponlop, 2009.
4 Maitripa, in Karma Chagmé, 1998: 78-80.
5 Klein, 2006: 171, 175.
6 Harvey, 2013: 341-2.
7 Sopa, 1978: 48-50, Wallace, 2006: 174-5.
8 Lalungpa, in Dakpo Tashi Namgyal, 2006: XXXVI-VIII.
9 Gyatrul Rinpoche, commentary to Padmasambhava, 1998: 108.
10 Harvey, 2013: 329.
11 Wallace, 2006: 123, 137-8.
12 Tsongkhapa, in Wallace 2005b: 218; Düdjom Lingpa, in Wallace 2011: 121; Buddhaghosa, 1999: 90-91.
13 Harvey, 2013: 92, 131, Wallace, 2006: 121-2, see also Sopa, 1978: 54-5.
14 Thrangu Rinpoche, 2004: 50, 212.
15 Buddhaghosa, 1999: 158, Tsongkhapa, 1997: 70.

16 Wallace, 2006: 13, Lodrö, 1998: 74, 221. The term introspection in this Buddhist meditation context is different from the term in Western psychology. In that setting it refers, in general terms, to examination of one's conscious thoughts and feelings.

17 Kabat-Zinn, 2005: 108, see also Segal, Williams, Teasdale, 2002: 40. Lutz et al., 2007: 509.

18 Analayo, 2003: 59, Bodhi, 2000: 76-9.

19 Buddhaghosa, 1999: 467, italics by avw; Harvey, 2013: 334-5.

Chapter 5 Three months of sitting with Settling

1 Some themes have been inspired by: Wallace, B. Alan, 2005a, *Genuine Happiness – Meditation as the Path to Fulfillment*, Hoboken, John Wiley and Sons, 22-25.

Chapter 6 Settling the mind and "tasting" the texts

1 Wallace 2011: X, XIV.

2 Lerab Lingpa (no publication date) *Commentary on the Dzogchen teachings called Heart Essence of Vimalamitra*. (The text by Lerab Lingpa is also presented in Wallace, B. Alan, 2006, *The Attention Revolution – unlocking the power of the focused mind*. Somerville MA, Wisdom Publications, 81-3).

3 Wallace, 2006: 86, Thrangu Rinpoche, 2004: 106.

4 Thrangu Rinpoche, 2004: 124.

5 Dakpo Tashi Namgyal, 2006: 160, Traleg Kyabgon Rinpoche, 2012: 46, Brown, 2006: 154-5.

6 Wallace, 2006: 90-93.

7 Düdjom Lingpa, *The Vajra Essence: From the Matrix of Pure Appearances and Primordial Consciousness, a Tantra on the Self-originating Nature of Existence*, translated under the guidance of Gyatrul Rinpoche by B. Alan Wallace, 2003, private edition, 25-6.

8 Based in Düdjom Lingpa, in Wallace, B. Alan, 2011, *Stilling the mind – shamatha teachings from Düdjom Lingpa's Vajra Essence*, Wisdom Publications, Boston MA; parts of pages 134-146.

Numbers as given in the text between parentheses have been included by avw, for easy reference.

9 Wallace, 2006: 107.
10 Wallace, 2006: 103-4.
11 Wallace, 2011: 128-132.
12 Padmasambhava, 1998: 99-102, including commentary by Gyatrul Rinpoche.
13 Thrangu Rinpoche, 2004: 216-17.
14 Welwood, Schireson, Holecek, 2012: 41.
15 Düdjom Lingpa, 2003: 39-40.

Chapter 7 Settling the mind, Mahamudra and Dzogchen

1 Wallace about the First Panchen Lama, 2006: 121; Düdjom Lingpa, 2003: 24, 29; Karma Chagmé, 1998: 80.
2 Thrangu Rinpoche, 2004: 9-10, 90.
3 Wallace, 2005a: 22; Wallace, 2006: 103.
4 Sogyal Rinpoche, 2002: 47-8.
5 Thrangu Rinpoche in Kongtrul, 2002: 145-6, Thrangu Rinpoche, 2004: 4-5.
6 Padmasambhava, 1998: 92.
7 Brown, 2006: 8-9.
8 Thrangu Rinpoche in Kongtrul, 2002: 126.
9 Brown, 2006: 151-152, 174.
10 Brown, 2006: 178-179.
11 Wallace, 2005b: 275, 305-8.
12 Brown, 2006: 14.
13 Thrangu Rinpoche in Kongtrul, 2002: 146.
14 Lutz et al., 2007: 514.
15 Padmasambhava, 1998: 114.
16 Traleg Kyabgon Rinpoche, 2012: 45.
17 Lutz et al., 2007: 511, 515.
18 Brown, 2006: 156.
19 Brown, 2006: 361, 364.
20 Norbu, 2000: 129-30.

21 Thrangu Rinpoche in Kongtrul, 2002: 142-3.
22 Thrangu Rinpoche, 2004: 21, 119.
23 Düdjom Lingpa, *The Vajra Essence: From the Matrix of Pure Appearances and Primordial Consciousness, a Tantra on the Self-originating Nature of Existence,* translated under the guidance of Gyatrul Rinpoche by B. Alan Wallace, 2003, private edition, 39.
24 Brown, 2006: xxiii.
25 Karma Chagmé, 1998: 111.
26 Dakpo Tashi Namgyal, 2006: 206.
27 Kongtrul, 2002: 65, Thrangu Rinpoche in Kongtrul, 2002: 139.
28 Brown, 2006: xvi-xxiv.
29 Palden, 2002.

Chapter 8 Contemplative and psychological views

1 For descriptions in this section, I feel nourished by Epstein, 1995: 15-41, Ray, 2000: 262-277, and Aronson, 2004.
2 Epstein, 1995: 30, Goleman, 2003: 61.
3 Ray, 2000: 267-8.
4 Ray, 2000: 272; Epstein, 1995: 34-5.
5 About peak experience: see for instance Maslow, 1968; about flow: Csikszentmihalyi, 1997, Nakamura and Csikszentmihalyi, 2005: 89-90.
6 Ray, 2000: 271.
7 From tape: Compassion meditation, May 2010. This meditation, and the one in Chapter 11 guided by avw, have been recorded in the Mind Centre of Phuket International Academy in Thailand. The remedies or guardians for compassion, as named, have been addressed by Alan Wallace, see Chapter 3.1.
8 About personality: see Sadock and Sadock, 2007: 304. The present description and some descriptions to follow in the context of psychiatry are based in this classic Western "Synopsis of Psychiatry." For DSM approach: see DSM-IV-TR (2000).
9 Epstein, 1995:18.

10 The book is by Phebe Cramer, 2006: 7; McCullough Vaillant, 1997: 25.

11 Described by Maslow as seeking to further a cause beyond the self (like service to others) and to experience a communion beyond the boundaries of the self through peak experience. This may involve mystical experiences and certain experiences with nature, aesthetic experience and/or transpersonal experiences. See: Koltko-Rivera, 2006: 302, and also: Maslow, 1968, Wilber, 2000: 212.

12 DSM-IV-TR Defensive Functioning Scale, 2000, in Sadock and Sadock, 2007: 312. On defensive functioning: see for instance Freud, Anna, 1936, Vaillant, 1992, McCullough Vaillant, 1997, Cramer, 2006, Van Waning, 2002, 2006, Sadock and Sadock, 2007: 792-4.

13 Wilber, in Wilber et al., 1986: 116-126, Walsh, 2009: 4-12.

14 Based in Chödrön, 1997: 65-72.

15 Freud, 2002.

16 For a general overview, including reference to Gebser and Maslow, see Wilber, 2006: 2-26, 54-69; Esbjörn-Hargens, 2010: 33-61.

Chapter 9 Psychological turbulence and self-healing

1 Panchen Lozang Chökyi Gyaltsen, in Wallace, 2009a: 57.

2 Some themes have been inspired by: Wallace, B. Alan, 2009a, *Mind in the Balance: meditation in science, Buddhism, and Christianity*. New York, Columbia University Press, 47-51.

3 Klein, 2006: 170-1, 335, n. 6; Clifford, 1984: 132-139.

4 With these drawings of yo-yoing and whiplash, there is a little section of implied time axis along the horizontal.

5 Lukoff et al., 1996: 243-4, Phillips et al., 2009: 68; see also De Waard, 2010.

6 Grof et al., 2008: 173-4.

7 Wallace, 2006: 102.

8 Singer, 2010: 167.

9 Almaas, 2002: 330.

10 Engler, 2003: 62-3, quoting Fingarette, 1963.

11 Almaas, 1988: 374, italics avw.

12 Jung, in Moacanin, 1986: 43; Freud, 1958: 135.

13 Clifford, 1984: 219-20 and 138-9.

14 Podvoll, 2003: xiii.

15 Ricard, 2003: 108-119; also Seligman, 2005: 3-9, Snyder and Lopez, 2005, Vaillant, 2008.

16 Wallace, 2011: 142.

17 Wallace, 2006: 102; Wallace, 2009a: 60-61.

Chapter 10 Contemplation and science, some musings

1 Sohlberg and Mateer, 2001: 7-8, 128-9.

2 Lutz et al., 2007: 503, 523-9; Lutz et al., 2008.

3 Wallace, 2006: 79; Siegel, 2007: 13; Teasdale et al., 2002.

4 For instance, including about mechanisms at work: Cranson et al., 1991, Alexander et al., 1994, Cahn and Polich, 2006: 200-2, Wilber, 2006: 11, 137.

5 Bartels and Zeki, 2004; see also Singer et al., 2004.

6 See Lutz et al. 2007: 543, for a description of various studies. For the Cultivating Emotional Balance research, see Kemeny et al., 2012.

7 Fraser, 2011: 33.

8 Some themes have been inspired by: Wallace, B. Alan, 2004, *The Four Immeasurables - Cultivating a Boundless Heart.* Ithaca NY, Snow Lion Publications, 135-8.

9 Research in this context, for instance: Brefczynski-Lewis et al., 2007, Lutz et al., 2007, 2008, 2009, Davidson and Lutz, 2008, Manna et al., 2010, Perlman et al., 2010 and Slagter et al., 2011. Description of the meditation styles: Lutz et al. 2007: 511-17.

10 To name a few recent accessible publications in this field: Begley, 2007, Siegel, 2007 (reference is made to: 190-1), 2010; Lutz et al., 2007, Goleman et al., 2008, Hanson and Mendius, 2009 (reference is made to: 52-5, 191-204), Austin, 2009.

11 To name a few: Hamilton-Merritt, 1997, Bucknell, 1997, Kang, 1997, Barendregt, 1996, Walsh, 1997, Dobisz, 2004, Siegel, 2007 and Siff, 2010.

12 Lutz et al., 2007: 519-521.

13 I am grateful to Daniel J.V. Beetsma for his statistical analyses and support.

Chapter 11 Applications and integration: shamatha in society

1 Hanson, 2008: 274, 280.

2 Gallagher and Zahavi, 2008: 34. About neurophenomenology: see Varela, 1996, and Lutz and Thompson, 2003.

3 Bitbol, 2008: 17-19.

4 Workshops are tailored to specific groups. They have been done with professionals in psychiatry and mindfulness fields. Workshops have also been offered to societal organizations like a bank foundation, active in social, sustainable environmental and cultural change, and EDE, Ecovillage Design Education, involved in designing sustainable settlements.

5 Zylowska, 2012. Biofeedback and shamatha: Jim Cahill, participant in the Shamatha Project, combines various biofeedback approaches with shamatha in Mindfulness Based Biofeedback Therapy (MBBT); Kees Blase and avw offer a combination of shamatha with Heart Rate Variability biofeedback, Blase and Van Waning, 2013.

6 Halifax, 2008: 45-6; Longaker, 2001.

7 Cohen, Leonard: 2012, lines from song nr 7: "Come Healing," on cd "Old Ideas."

8 Van Lommel, 2010: 46; Moorjani, 2012: 163, 166.

9 Wallace, 2006: 122; Wallace, 2009a: 92-4. Italics avw.

10 Wallace and Hodel, 2008: 193, quoting Henning Genz, 1999: 26; Wallace, 2006: 81.

Chapter 12 On human flourishing: continuing the project

1 Some themes have been inspired by: Wallace, B. Alan, 2006, *The*

Attention Revolution – unlocking the power of the focused mind. Somerville MA, Wisdom Publications, 58.

2 Wallace, 2005a: 113.

3 Jacobs et al., 2011: 676. In this section I return to some findings as referred to in Chapter 1: for these and all described outcomes, see Shamatha Project, in References section.

4 *Nadi* (Sanskrit): the internal channels of the subtle body (see also Ch 4.1). Welwood, Schireson and Holecek, 2012: 37.

5 For more information, and the press release "Templeton Foundation awards grant for meditation research (VIDEO)," including naming the co-investigators and trainees on the grant team, and all the Funds that have supported the work since 2006: see References, Shamatha Project, UC Davis News and Information, November 19th, 2012.

6 International Shamatha Project, web link.

7 Thrangu Rinpoche, 2004: 21; Thrangu Rinpoche in Kongtrul, 2002: 139; Padmasambhava, 1998: 92.

References

Alexander, Charles N., Heaton, D.P., and Chandler, H.M., 1994, "Advanced Human Development in the Vedic Psychology of Maharishi Mahesh Yogi: Theory and Research." In: Miller, M.E. and Cook-Greuter, S.R. (eds.), *Transcendence and Mature Thought in Adulthood – the further reaches of adult development*, Lanham MD, Rowman and Littlefield, 39-70.

Almaas, A.H., 1988, *The Pearl Beyond Price - Integration of Personality into Being: an Object Relations Approach*, Boston MA, Shambhala.

Almaas, A.H., 2002, *Spacecruiser Inquiry – True guidance for the Inner Journey*, Boston MA, Shambhala.

Analayo, 2003, *Satipatthana, The Direct Path to Realization*, Birmingham, Windhorse.

Anderson, Sherry R. and Hopkins, Patricia, 1991, *The Feminine Face of God - the unfolding of the sacred in women*, New York, Bantam.

Aronson, Harvey, 2004, *Buddhist practice on Western ground – reconciling Eastern ideals and Western psychology*, Boston MA, Shambhala.

Association for Psychological Science, 2010, 14 July, *Association for Psychological Science Press release*, available from: http://www.psychologicalscience.org/index.php/news/releases/meditation-helps-increase-attention-span.html (accessed 1 April 2013).

Austin, James H., 2009, *Selfless insight: Zen and the meditative transformations of consciousness*, Cambridge MA, Massachusetts Institute of Technology Press.

"Ayacana Sutta – The Request," trans. Thanissaro Bhikkhu, 1997 (SN 6:1, S i 136), *Access to Insight*, available from http://www.accesstoinsight.org/tipitaka/sn/sn06/sn06.001.than.html (accessed 1 April 2013).

Barendregt, Henk P., 1996, "Mysticism and beyond, Buddhist phenomenology Part II," *The Eastern Buddhist, New Series*, 29, 262-287, available from http://www.cs.ru.nl/~henk/BP/bp2.html

(accessed 1 April 2013).

Bartels, Andreas, and Zeki, Semir, 2004, "The neural correlates of maternal and romantic love," *NeuroImage,* 21(3), 1155-1166.

Begley, Sharon, 2007, *Train Your Mind, Change Your Brain – a ground-breaking collaboration between neuroscience and Buddhism,* New York, Ballantine.

Bitbol, Michel, 2008, "Is consciousness primary?" *NeuroQuantology,* 2008, vol. 6, nr 1, 53-72.

Blase, Kees and Van Waning, Adeline, 2013, "Meditation at the monitor, Heart Rate Variability and cortisol during shamatha meditation," to be published.

Bodhi, Bhikkhu (ed.), 2000, *Abhidhammattha Sangaha of Acariya Anuruddha, a Comprehensive Manual of Abhidhamma,* Seattle, BPS Buddhist Publication Society Pariyatti Edition.

Bodhi, Bhikkhu (ed., introd.), 2005, "Kalama Sutta – To the Kalamas" (AN 3: 65, A I 188), in Ch. III "Approaching the Dhamma," *In the Buddha's Words – An Anthology of Discourses from the Pali Canon,* Boston, Wisdom, 89-90.

Bond, Michael, 2011, "Everybody Say OM – meditation isn't just for mystics, it can actually improve your mental and physical health," *New Scientist,* Jan. 8, 32-35.

Brefczynski-Lewis, Julie A., Lutz, A., Schaefer, H.S., Levinson, D.B., and Davidson, R.J., 2007, "Neural correlates of attentional expertise in long-term meditation practitioners," *PNAS,* vol. 104 (27), 11483-11488.

Brown, Daniel P., 2006, *Pointing out the Great Way, the stages of meditation in the Mahamudra tradition,* Boston MA, Wisdom.

Bucknell, Rod, 1997, "Experiments in Insight Meditation," in Bucknell, R. and Kang, C. (eds.), *The Meditative Way – Readings in the theory and practice of Buddhist meditation,* Richmond UK, Curzon, 244-263.

Buddhaghosa, 1999, *Visuddhimagga, The Path of Purification,* Nanamoli (trans.), Kandy, Sri Lanka, Pariyatti Publishing /Buddhist Publication Society.

Cahn, B. Rael and Polich, John 2006, "Meditation states and traits: EEG, ERP, and neuroimaging studies," *Psychological Bulletin*, 132, 180-211.

Chödrön, Pema, 1997, *When things fall apart: heart advice for difficult times*, Boston MA, Shambhala.

Clifford, Terry, 1984, *Tibetan Buddhist Medicine and Psychiatry – The Diamond Healing*, York Beach ME, Samuel Weiser.

Cohen, Leonard, 2012. *Old Ideas*, music-cd. LLC, Little Fountain Music, Columbia.

Cramer, Phebe, 2006, *Protecting the self: Defense Mechanisms in action*, New York, Guilford.

Cranson, Robert W., Orme-Johnson, D.W., Dillbeck, M.C., Jones, C.H., Alexander, C.N., and Gackenback, J., 1991, "Transcendental Meditation and improved performance on intelligence-related measures: A longitudinal study," *Journal of Personality and Individual Differences*, 12, 1105-16.

Csikszentmihalyi, Mihaly, 1997, *Finding Flow: the Psychology of Engagement with Everyday Life*, New York, Basic Books.

Dakpo Tashi Namgyal, 2006, *Mahamudra – The moonlight – quintessence of mind and meditation*, trans. and ann. by Lobsang P. Lhalungpa, Boston MA, Wisdom.

Davidson, Richard and Lutz, Antoine, 2008, "Buddha's Brain: Neuroplasticity and Meditation," *IEEE Signal Process Mag.*, 25 (1), 176-184.

De Waard, Fransje, 2010, *Spiritual Crisis – Varieties and Perspectives of a Transpersonal Phenomenon*, Exeter UK, Imprint Academic.

Dobisz, Jane, 2004, *The Wisdom of Solitude – A Zen Retreat in the Woods*, San Francisco CA, Harper San Francisco.

DSM-IV-TR, 2000, Diagnostic and Statistical Manual of Mental Disorders, 4th ed. Text revision, Washington DC, American Psychiatric Association.

DSM-IV-TR "Defensive Functioning Scale," in Sadock, Benjamin J. and Sadock, Virginia A., 2007, *Synopsis of psychiatry – behavioural sciences / clinical psychiatry*, Baltimore MA, Williams and Wilkins,

312.

Düdjom Lingpa, 2003, *The Vajra Essence – from the Matrix of Pure Appearances and Primordial Consciousness, a Tantra on the Self-originating Nature of Existence,* trans. B.A. Wallace, guidance Gyatrul Rinpoche, private edition.

Ekman, Paul, 2007, *Emotions revealed: Recognizing Faces and Feelings to improve Communication and Emotional Life,* New York, Owl Books.

Ekman, Paul, and Friesen, W., 1978, *Facial Action Coding System: A Technique for the Measurement of Facial Movement,* Palo Alto CA, Consulting Psychologists Press.

Engler, Jack, 2003, "Being somebody and being nobody: a re-examination of the understanding of self in psychoanalysis and Buddhism," in Safran, J.D. (ed.) *Psychoanalysis and Buddhism: an unfolding dialogue,* Somerville MA, Wisdom, 35-100.

Epel, Elissa, Daubenmier, J., Moskowitz, J.T., Folkman, S., and Blackburn, E., 2009, "Can meditation slow rate of cellular aging? Cognitive stress, mindfulness, and telomeres," *Annals of the New York Academy of Sciences,* 1172: 34-53.

Epstein, Mark, 1995, *Thoughts without a Thinker: Psychotherapy from a Buddhist Perspective,* New York, Basic Books.

Esbjörn-Hargens, Sean, 2010, "An Overview of Integral Theory – an all-inclusive framework for the twenty-first century," in Esbjörn-Hargens, S. (ed.) *Integral Theory in Action – applied, theoretical, and constructive perspectives on the AQAL model,* Albany NY, SUNY Press, 33-61, available from http://www.integrallife.com/node/37539 (accessed 1 April 2013).

Fingarette, Herbert, 1963, *The Self in Transformation,* New York, Basic Books.

Fraser, Andy, 2011, "The Shamatha Project," *View,* July, 28-33.

Fraser, Andy, 2013, *The healing power of meditation: leading experts on Buddhism, psychology, and medicine explore the health benefits of contemplative practice,* London, UK, Random House.

Freud, Anna, 1936, *The Ego and the Mechanisms of Defense,* London UK, Karnac Books.

Freud, Sigmund, 2002, *Civilization and its discontents*, London UK, Penguin (orig. 1930).

Freud, Sigmund, 1958, "On Beginning the Treatment (Further Recommendations on the Technique of Psychoanalysis)," vol. 12, *Standard Edition of the Complete Psychological Works of Sigmund Freud*, ed. and trans. James Strachey, London UK, Hogarth Press, 135 (orig. 1913).

Gallagher, Shaun, and Zahavi, Dan, 2008, *The Phenomenological Mind – an introduction to philosophy of mind and cognitive science*, New York, Routledge.

Genz, Henning, 1999, *Nothingness: The Science of Empty Space*, Cambridge MA, Perseus.

Gethin, Rupert, 1998, *The Foundations of Buddhism*, Oxford UK, Oxford University Press.

Goleman, Daniel, 2003, *Destructive Emotions – How can we overcome them? A Scientific Dialogue with the Dalai Lama*, New York, Bantam.

Goleman, Daniel, and others, 2008, *Measuring the Immeasurable – The Scientific Case for Spirituality*, Boulder CO, Sounds True.

Grof, Stan, Grob, C., Bravo, G., and Walsh, R., 2008, "Birthing the Transpersonal," *Journal of Transpersonal psychology*, 115-177.

Halifax, Joan, 2008, *Being with Dying – cultivating compassion and fearlessness in the presence of death*, Boston MA, Shambhala.

Hamilton-Merritt, 1997, "From a Meditator's Diary," in Bucknell, R. and Kang, C. (eds.), *The Meditative Way – Readings in the theory and practice of Buddhist meditation*, Richmond UK, Curzon Press, 208-218.

Hanson, Rick, 2008, "Seven facts about the brain that incline the mind to joy," in Goleman, D., and others, *Measuring the Immeasurable – The Scientific Case for Spirituality*, Boulder CO, Sounds True, 269-286.

Hanson, Rick, and Mendius, Richard, 2009, *Buddha's Brain – the Practical Neuroscience of happiness, love and wisdom*, Oakland CA, New Harbinger.

Harvey, Peter, 2013, *An introduction to Buddhism – teachings, history and practices,* Second Edition, Cambridge UK, Cambridge University Press.

Ingersoll, R. Elliot and Zeitler, David M. 2010, *Integral Psychotherapy – Inside Out/Outside In,* Albany NY, SUNY Press.

Jacobs, Tonya L., Epel, E.S., Lin, J., Blackburn, E.H., Wolkowitz, O.M., Bridwell, D.A., Zanesco, A.P., Aichele, S.R., Sahdra, B.K., MacLean, K.A., King, B.G., Shaver, P.R., Rosenberg, E.L., Ferrer, E., Wallace, B.A., and Saron, C.D., 2011, "Intensive meditation training, immune cell telomerase activity, and psychological mediators," *Psychoneuroendocrinology,* 36 (5), 664-81.

Jacobs, Tonya L., Shaver P.R., Epel, E.S., Zanesco, A.P., Aichele, S.R., Bridwell, D.A., King, B.G., MacLean, K.A., Ferrer, E., Rosenberg, E.L., Sahdra, B.K., Kemeny, M.E., Wallace, B.A., and Saron, C.D., 2013, "Self-Reported Mindfulness and cortisol during a Shamatha meditation retreat," *Health Psychology* (in press).

Jung, Carl G., "Commentary on the Secret of the Golden Flower," in: Moacanin, R., *Jung's Psychology and Tibetan Buddhism,* 1986, Boston MA, Wisdom.

Kabat-Zinn, Jon, 2005, *Coming to our Senses – healing ourselves and the world through mindfulness,* New York, Hyperion.

"Kalama Sutta – To the Kalamas," 2005, (AN 3: 65, A I 188), in Bodhi, Bhikkhu (ed., introd.), Ch. III "Approaching the Dhamma," *In the Buddha's Words – An Anthology of Discourses from the Pali Canon,* Boston MA, Wisdom, 89-90.

Kang, Chris, 1997, "Experiences in Meditation," in Bucknell, R. and Kang, C. (eds.), *The Meditative Way – Readings in the theory and practice of Buddhist meditation,* Richmond UK, Curzon Press, 197-207.

Karma Chagmé, 1998, *A Spacious Path to Freedom,* comm. Gyatrul Rinpoche, transl. B. A. Wallace, Ithaca NY, Snow Lion.

Kemeny, Margaret, Folts, C., Cavannagh, J.F., Culen, M., Giese-Davis, J., Jennings, P., Rosenberg, E.L., Gilath, O., Shaver, P.R., and Ekman, P., 2012, "Contemplative/emotion training reduces negative emotional behavior and promotes prosocial responses," *Emotion*, April 12 (2): 338-50.

Klein, Anne C., 2006, "Psychology, the Sacred, and Energetic Sensing," in Unno, M. (ed.), *Buddhism and Psychotherapy across cultures – essays on theories and practices*, Somerville MA, Wisdom, 169-192.

Koltko-Rivera, Mark E., 2006, "Rediscovering the later version of Maslow's Hierarchy of Needs: self-transcendence and opportunities for theory, research, and unification," *Review of General Psychology*, 10 (4), 302-317.

Kongtrul Jamgön, 2002, *Creation and completion, Essential points of tantric meditation*, transl. and introd. Harding, S., comm. Khenchen Thrangu Rinpoche, Somerville MA, Wisdom.

LaBerge, Stephen, 2003, "Lucid Dreaming and the Yoga of the Dream State: a psychophysiological perspective," in Wallace, B.A. (ed.) *Buddhism & Science – breaking new ground*, New York, Columbia University Press, 233-258.

Lau, D.C. (transl.) 1979. *Confucius: The Analects*, New York, Penguin.

Lerab Lingpa, (no publication date), *Commentary on the Dzogchen teachings called Heart Essence of Vimalamitra*, written down by Taklung Tsetrul Pema Wangyal (ed.), 638-640.

Lodrö, Geshe Gedün, 1998, *Calm Abiding and Special Insight*, transl. and ed. by Jeffrey Hopkins, Ithaca NY, Snow Lion.

Longaker, Christine, 2001, *Facing Death and Finding Hope – a guide to the emotional and spiritual care of the dying*, New York, Broadway Books.

Lukoff, David, Lu, F.G., and Turner, R., 1996, "Diagnosis: A Transpersonal Clinical Approach to Religious and Spiritual Problems," in Scotton, B.W., Chinen, A.B., Battista, J.R. (eds.), *Textbook of Transpersonal Psychiatry and Psychology*, New York, Basic Books, 231-249.

Lutz, Antoine and Thompson, Evan, 2003, "Neurophenomenology – Integrating Subjective Experience and Brain Dynamics in the Neuroscience of Consciousness," *Journal of Consciousness Studies*,

10, (9-10), 31-52.

Lutz, Antoine, Dunne, J.D. and Davidson, R.J., 2007, "Meditation and the Neuroscience of Consciousness: An Introduction," in Zelazo, P.D., Moscovitch, M. and Thompson, E. (eds.), *The Cambridge Handbook of Consciousness*, Cambridge, Cambridge University Press, 499-551.

Lutz, Antoine, Slagter, H.A., Rawlings, N.B., Francis, A.D., Greischar, L.L., and Davidson, R., 2009, "Mental Training enhances attentional stability: neural and behavioral evidence," *The Journal of Neuroscience*, 29 (42), 13418-13427.

MacLean, Katherine, Ferrer, E., Aichele, S.R., Bridwell, D.A., Zanesco, A.P., Jacobs, T.L., King, B.G., Rosenberg, E.L., Sahdra, B.K., Shaver, P.R., Wallace, B.A., Mangun, G.R., and Saron, C.D., 2010, "Intensive meditation Training Improves Perceptual Discrimination and Sustained Attention," *Psychological Science*, June, 21, 829-39.

Manna Antonietta, Raffone, A., Perrucci, M.G, Nardo D., Ferretti, A. Tartaro, A., Londei, A., Del Gratta, C., Belardinelli M.O., and Romani G.L, 2010, "Neural correlates of focused attention and cognitive monitoring in meditation," *Brain Research Bulletin*, April 29; 82 (1-2); 46-56.

Marchant, Jo, 2011, "How meditation might ward off the effects of ageing," *The Observer*, April 24.

Maslow, Abraham, 1968, "The farther reaches of human nature," *Journal of Transpersonal Psychology*, 1 (1), 1-9.

McCullough Vaillant, Leigh, 1997, *Changing character: short-term anxiety-regulating psychotherapy for restructuring defenses, affects, and attachment*, New York, Basic Books.

Moorjani, Anita, 2012, *Dying to be me – my journey from cancer, to Near Death to True Healing*, London UK, Hay House.

Nakamura, Jeanne and Csikszentmihalyi, Mihaly, 2005, "The Concept of Flow," in Snyder, C.R., and Lopez, S.J. (eds.), *Handbook of Positive Psychology*, Oxford UK, Oxford University Press, 89-105.

Norbu, Namkhai, 2000, *The Crystal and the Way of Light – Sutra, Tantra and Dzogchen*, comp. and ed. Shane, J., Ithaca NY, Snow Lion.

Nyoshul Khenpo Rinpoche (no date), Patrick Gaffney, *Nyoshul Khen Rinpoche,* available from Ranjung Yeshe Publications http://www.rangjung.com/authors/Nyoshul_Khen_Rinpoche_tribute.htm (accessed 1 April 2013).

Padmasambhava, 1998, *Natural Liberation – Padmasambhava's teachings on the six Bardos*, comm. Gyatrul Rinpoche, trans. B. A. Wallace, Boston MA, Wisdom.

Palden Drolma, Lama, 2002, "Working with the Western mind," *Mandala Magazine*, June-August.

Perlman, David M., Salomons T.V., Davidson J.J., and Lutz, A, 2010, "Differential effects on pain intensity and unpleasantness of two meditation practices," *Emotion*, 10 (1), 65-71.

Perry, Whitall (ed.), 1981, *The Treasury of Traditional Wisdom*, Pates, Middlesex UK, Perennial Books.

Phillips, Russell E., Lukoff, D., and Stone, M.K., 2009, "Integrating the Spirit within Psychosis: Alternative conceptualizations of psychotic disorders," *The Journal of Transpersonal Psychology*, 41, 61-80.

Podvoll, Edward M., 2003, *Recovering Sanity – a compassionate approach to understanding and treating psychosis*, Boston MA, Shambhala.

Ponlop, Dzogchen, 2009, "Pointing out Ordinary Mind," *Buddhadharma – The Practitioners quarterly*, Fall.

Ray, Reginald A., 2000, *Indestructible Truth, the living spirituality of Tibetan Buddhism*, Boston MA, Shambhala.

Ricard, Matthieu, 2003, *Happiness: a guide to developing life's most important skill*, New York, Little, Brown and Company.

Sadock, Benjamin J. and Sadock, Virginia A., 2007, *Synopsis of psychiatry – behavioural sciences/clinical psychiatry*, Baltimore MA, Williams and Wilkins.

Sahdra, Baljinder K., MacLean, K.A., Ferrer, E., Shaver, P.R.,

Rosenberg, E.L., Jacobs, T.L., Zanesco, A.P., King, B.G., Aichele, S.R., Bridwell, D.A., Mangun, G.R., Lavy, S., Wallace, B.A., and Saron, C.D., 2011, "Enhanced Response Inhibition During Intensive Meditation Training Predicts Improvements in Self-Reported Adaptive Socioemotional Functioning," *Emotion,* 11(2), 299-312.

Saggar, Manish, King, B.G., Zanesco, A.P., MacLean, K.A., Aichele, S.R., Jacobs, T.L., Bridwell, D.A., Shaver, P.R., Rosenberg, E.L., Sahdra, B.K., Ferrer, E., Tang, A.C., Mangun, G.R., Wallace, B.A., Miikkulainen, R., and Saron, C.D., 2012, "Intensive training induces longitudinal changes in meditation state-related EEG oscillatory activity," *Frontiers in Human Neuroscience,* 10 September, doi: 10.339/fnhum.2012.00256, 1-14.

Santa Barbara Institute of Consciousness Studies, SBI, with B. Alan Wallace PhD as Director; including audio-recordings of meditation guidance, http://www.sbinstitute.com (accessed 4 April 2013).

Saron, Clifford D., 2009, "Results of the Shamatha Project," Mind and Life XVIII, Dialogues between Buddhism and the Sciences, "Attention, Memory and Mind – a synergy of psychological, neuroscientific, and contemplative perspectives," with His Holiness the Dalai Lama, Dharamsala, India, April 6-10, 2009, available from http://www.mindandlife.org/Dharamsala-Apr6-2009Brochure.pdf , p. 9 (accessed 30 May 2009).

Saron, Clifford D., 2013, "Training the mind: The Shamatha Project," in: Fraser, A. (ed.) *The healing power of meditation: leading experts on Buddhism, psychology, and medicine explore the health benefits of contemplative practic,* London UK, Random House, 45-65.

Segal, Zindel, Williams, J. M.G., Teasdale, J.D., 2002, *Mindfulness-Based Cognitive Therapy for Depression – a new approach to preventing relapse,* New York, Guilford.

Seligman, Martin E.P., 2005, "Positive Psychology, Positive Prevention, and Positive Therapy," in Snyder, C.R., and Lopez, S.J. (eds.), *Handbook of Positive Psychology,* Oxford UK, Oxford

University Press, 3-9.

Shah, Idres, 1971, *The Sufi*, Garden City NY, Anchor/Doubleday.

Shamatha Project, The, see: *UC Davis*, Shamatha Project, Saron Lab, http://mindbrain.ucdavis.edu/labs/Saron/shamatha-project (accessed 1 April 2013), for an overview of the project, including information about the investigators and consulting scientists, funding and support, and for up to date outcomes of the Shamatha Project, including the complete articles-texts in pdf. format of scientific research outcomes; see also: *Santa Barbara Institute for Consciousness Studies*, CA, for information and weblinks http://www.sbinstitute.com/Shamatha_Project (accessed 1 April 2013).

Shamatha Project, The, 2012, "Templeton Foundation awards grant for meditation research (VIDEO)," *UC Davis News and Information*, 19 November, available at http://news.ucdavis.edu/search/news_detail.lasso?id=10420 (accessed 1 April 2013) (video: Clifford Saron on TEDx).

Shamatha Project, The International, Santa Barbara Institute for Consciousness Studies, CA, available at http://www.shamatha.org (accessed 1 April 2013).

Siegel, Daniel J., 2007, *The Mindful Brain – Reflection and Attunement in the cultivation of well-being*, New York, Norton and Company.

Siegel, Daniel J., 2010, *The Mindful Therapist – A clinician's guide to Mindsight and Neural Integration*, New York, Norton and Company.

Siff, Jason, 2010, *Unlearning Meditation – What to do when the instructions get in the way*, Boston MA, Shambhala.

Singer, Tania, Seymour, B., O'Doherty, J., Kaube, H., Dolan, R.J., and Frith, C.D., 2004, "Empathy for pain involves the affective but not sensory components of pain," *Science*, 303 (5661), 1157-62.

Singer, Thea, 2010, *Stress Less (for women)*, New York, Plume-Penguin.

Slagter, Heleen A., Davidson, R.J., and Lutz, A., 2011, "Mental training as a tool in the neuroscientific study of brain and cognitive plasticity," *Frontiers in Human Neuroscience*, 01/2011; 5: 17.

Snyder, Charles R., and Lopez, S.J. (eds.), 2005, *Handbook of Positive Psychology*, Oxford UK, Oxford University Press.

Sogyal Rinpoche, 2002, *The Tibetan Book of Living and Dying*, New York, Harper Collins.

Sohlberg, McKay Moore, and Mateer, Catherine A., 2001, *Cognitive Rehabilitation: An Integrative Neuropsychological Approach*, New York, Guilford.

Sopa, Geshe, 1978, "*Samathavipasyanayuganaddha*: The Two Leading Principles of Buddhist Meditation," in Kiyota, M. (ed.) *Mahayana Buddhist Meditation: Theory and Practice*, Honolulu, University Press of Hawaï, and Delhi, Motilal Banarsidass (1998), 46-65.

Tarthang Tulku, 2003, *Tibetan Relaxation – Kum Nye Massage and Movement, a yoga for healing and energy from the Tibetan tradition*, Berkeley, CA, Dharma Publishing.

Tarthang Tulku, 2012, *Kum Nye Dancing – Introducing the mind to the treasures the body offers*, Berkeley, CA, Dharma Publishing.

Teasdale, John D., Moore, R.G., Hayhurst, H., Pope, M., Williams, S., Segal, Z.V., 2002, "Metacognitive awareness and prevention of relapse in depression: empirical evidence," *J. of Consulting and Clinical Psychology*, 70, 275-87.

Tenzin Wangyal Rinpoche, 1998, *The Tibetan yogas of dream and sleep*, Ithaca NY, Snow Lion.

Thanissaro, Bhikkhu, (trans.), 1997, "Ayacana Sutta – The Request," (SN 6:1, S i 136), *Access to Insigh*t, available from http://www.accesstoinsight.org/tipitaka/sn/sn06/sn06.001.than.html (accessed 1 April 2013).

Thich Nhat Hanh, 1998, *The Heart of the Buddha's Teaching*, Berkeley CA, Parallax Press.

Thrangu, Khenchen Rinpoche, 2004, *Essentials of Mahamudra: looking directly at the mind*, Somerville MA, Wisdom.

Traleg Kyabgon Rinpoche, 2012, "Meditating on the mind itself," *Buddhadharma – The Practitioners Quarterly*, Winter, 44-47.

Tsongkhapa, 1997, "Fastening the Mind," in Bucknell, R. and Kang, C. (eds.) *The Meditative Way – Readings in the theory and practice of Buddhist meditation*, Richmond UK, Curzon Press, 69-77.

Vaillant, George E., 1992, *Ego mechanisms of defense: A guide for clinicians and researchers*, Washington DC, American Psychological Association.

Vaillant, George E., 2008, *Spiritual Evolution – a scientific defense of faith*, New York, Broadway Books.

Van Lommel, Pim, 2010, *Consciousness Beyond Life – The Science of the Near-Death Experience*, New York, Harper Collins.

Van Waning, Adeline, 2002, "A mindful self and beyond – sharing in the ongoing dialogue of Buddhism and psychoanalysis," in Young-Eisendrath, P., Muramoto, S. (eds.) *Awakening and Insight: Buddhism and Psychotherapy East and West*, New York, Brunner Routledge, 93-105.

Van Waning, Adeline, 2006, "Buddhist Psychology and Defensive Conditioning," in Kwee, M.G.T., Gergen, K.J. and Koshikawa, F. (eds.) *Horizons in Buddhist Psychology: Practice, Research & Theory*, Taos NM, Taos Institute Publishing, 141-153.

Van Waning, Adeline, 2011, "Inside the Shamatha Project," *Buddhadharma – The Practitioner's Quarterly*, Summer, 39-45.

Varela, Francisco J., 1996, "Neurophenomenology: a methodological remedy to the hard problem," *Journal of Consciousness Studies*, vol. 3, 330-350.

Wallace, B. Alan, 2004, *The Four Immeasurables – Cultivating a Boundless Heart*, Ithaca NY, Snow Lion.

Wallace, B. Alan, 2005a, *Genuine Happiness – Meditation as the Path to Fulfillment*, Hoboken NJ, John Wiley and Sons.

Wallace, B. Alan, 2005b, *Balancing the Mind – a Tibetan Buddhist Approach to Refining Attention*, Ithaca NY, Snow Lion.

Wallace, B. Alan, 2006, *The Attention Revolution – unlocking the power of the focused mind*, Somerville MA, Wisdom.

Wallace, B. Alan, 2009a, *Mind in the Balance: meditation in science, Buddhism, and Christianity*, New York, Columbia University

Press.

Wallace, B. Alan, 2009b, "Within you, without you," *Tricycle*, Winter, 79-83, 119.

Wallace, B. Alan, 2011, *Stilling the mind – Shamatha teachings from Düdjom Lingpa's Vajra Essence*, Boston MA, Wisdom.

Wallace, B. Alan, 2012, *Dreaming yourself awake – lucid dreaming and Tibetan dream yoga for insight and transformation*, Boston MA, Shambhala.

Wallace, B. Alan and Hodel, Brian, 2008, *Embracing Mind – the common ground of science & spirituality*, Boston MA, Shambhala.

Walsh, Roger, 1997, "Initial Meditative Experiences," in Bucknell, R. and Kang, C. (eds.), *The Meditative Way – Readings in the theory and practice of Buddhist meditation*, Richmond UK, Curzon Press, 228-243 (orig. 1977).

Walsh, Roger, 1999, *Essential Spirituality - The 7 Central Practices to Awaken Heart and Mind*, New York, Wiley.

Walsh, Roger, 2009, "The state of the Integral Enterprise, Part I: Current Status and Potential Traps," *Journal of Integral Theory and Practice*, 4 (3), 1-12.

Welwood, John, Schireson, G., and Holecek, A., 2012, "Heal the self, free the self – bringing together Western psychology and Buddhism," *Buddhadharma – The Practitioners's Quarterly*, Summer edition, 32-41.

Wilber, Ken, 2000, *Integral Psychology – consciousness, spirit, psychology, therapy*, Boston MA, Shambhala.

Wilber, Ken, 2006, *Integral Spirituality: a startling new role for religion in the modern and postmodern world*, Boston MA, Shambhala.

Wilber, Ken, Engler, J. and Brown, D.P., 1986, *Transformations of Consciousness – conventional and contemplative perspectives on development*, Boston MA, Shambhala.

Zylowska, Lidia, 2012, *The Mindfulness prescription for adult ADHD*, Boston MA, Shambhala.

Select Index

Adeline van Waning lives and works mainly in The Netherlands. She gives meditation guidance, works in hospice care, and paints, www.meditationapr.org

MANTRA
BOOKS